Praise for *Mind O...* W9-CCB-761

"An absolute delight...Belongs on the bedside bookshelf of every science enthusiast and should also be a treat for any reader curious about the universe." —*San Jose Mercury-News*

"In pithy, three-page bursts, [Cole] tackles particle physics, geometry, Alfred Nobel, and about a thousand other topics, all with graceful, accessible prose." —*The Boston Globe*

"Quick, insightful, and often witty essays." —*Science News*

"If John Donne were alive today and living in Los Angeles, he would be a K.C. Cole fan. Her science columns have the elegance and concision of metaphysical poetry and sometimes the bite and pathos as well."
 —Jack Miles, Pulitzer Prize winner and MacArthur Fellow

"This book should be read by anyone interested in taking a stroll through the cosmos, stopping to admire new ideas, and appreciating our place in this universe."
 —Wendy Freedman, director of the Carnegie Observatories

"This fascinating and coherent collection captures much of life—from the deepest questions about the nature of our universe and the Big Bang to thong bikinis, the Nobel Prize, and mountains on Mars." —Andrew Strominger,
professor of physics, Harvard University

"K. C. Cole is a bright star in the science-writing universe."
 —Timothy Ferris, author of *Seeing in the Dark*

"To communicate the essence of science is a challenge, and K.C. Cole succeeds superbly, time after time illuminating key concepts for anyone with an inquiring mind."
 —Martin Rees, Astronomer Royal

MIND
OVER
MATTER

MIND

OVER

MATTER

Conversations with the Cosmos

K. C. Cole

A HARVEST BOOK • HARCOURT, INC.

Orlando   Austin   New York   San Diego   Toronto   London

Substantially different versions of "Mind Over Matter" and
"Objectivity" first appeared in *Discover* magazine.
"Play, by Definition, Suspends the Rules" Copyright © 1998
by the New York Times Co. Reprinted by permission.

Library of Congress Cataloging-in-Publication Data
Cole, K. C.
Mind over matter: conversations with the cosmos/K. C. Cole—1st ed.
p.   cm.
Includes index.
ISBN 0-15-100816-7
ISBN 0-15-602956-1 (pbk.)
1. Science—Popular works. I. Title.
Q162.C584 2003
500—dc21   2003000982

Text set in Garamond MT
Designed by Cathy Riggs

Printed in the United States of America

First Harvest edition 2004
A C E G I K J H F D B

For Walter,
And all the readers
who kept me minding
what matters.

Contents

PART II: STUFF

PART III: DOING IT

PART IV: "POLITICAL" SCIENCE

Introduction
and Acknowledgments

There's no escaping it: We live at the bottom of a gravity well. It's our local pothole in spacetime, created by the sagging weight of the Earth in the four-dimensional fabric that serves (as far as we know) as the backdrop for our universe. Stuck as we are, it's a strain to catch glimmers of what's beyond. We turn our shiny mirrored telescopes and radio antennas toward the sky, like upside-down umbrellas, hoping for a drop of news. We bury arrays of light catchers and particle traps deep in the ground, in the ocean, in the ice, waiting for messages in bottles written in some strange and indecipherable tongue.

What big eyes we have! What big ears!

We've even managed to boost an occasional probe up and out of the well and into the local neighborhood, an emissary to the world beyond. But we haven't gotten far, not even beyond our local star.

Thank goodness, the universe is a chatty place. It has an expansive vocabulary, an impressive range. It whispers to us in microwaves, shrieks in X-rays, rains down tales in the infrared, sparkles in visible light, grumbles in gravity waves, pounds us with cosmic rays, and slices right through us with neutrinos. Sometimes it teases, tugging on us with mysterious dark matter, sending down an occasion giggle of gamma rays. Each signal adds a different kind of clue.

It speaks not only of the great "out there" but also of the great "back when," because peering out into the universe is also peering back in time. We can look ourselves in the face at a time long before we were born, before there were planets, stars, even atoms.

The writing is on the wall, but it's often difficult to read. Not surprisingly, our conversation with the cosmos involves a great deal of translation, for as many languages as humans speak, nature speaks more and different ones. And so we commune in equations, metaphors, images, logic, emotions—whatever seems to work.

As in any ongoing discourse, there are (occasionally comical—at least in retrospect) misunderstandings. There are also moments of genuine, astonishing, connection. Unbelievable luck.

The translation problem is compounded by the fact that we are so tightly entangled with our subject, such intimate partners, the cosmos and us. The universe makes us in its image; we return the favor, making it in ours. It's not always easy to separate what is "out there" from what is "in us" or even to know to what extent these are useful distinctions. We are what we are because the universe has made atoms and space and time just so; we look back at what made us, and see, to some extent, ourselves.

Still, we can't help trying. We can no more resist the urge to get to know the universe than a baby can resist the lure of its own fingers and toes. This is who we are. This is where we came from, what we will become. Home, sweet home.

For the past seven years, I've been blessed with the opportunity to carry on an ongoing, public discourse about the universe, the stuff it's made of, the people who explore it (or just plain admire it), in a column in the pages of the *Los Angeles Times,* called "Mind Over Matter." Those essays are revised and collected here, along with a few earlier columns from *Discover* and the *New York Times.* Although there is no separating the business of seeing from what is seen, or what is seen from who is doing the looking (and how), I've made the following attempt:

Section I, "The I of the Beholder," focuses on the central role of us, the seers and seekers, and our relationship with what is seen. Section II, "Stuff," focuses on what's "out there." Section III, "Doing It," focuses on some of the people who converse with the cosmos, how they do it, the obstacles they encounter. Section IV, "Political Science," focuses on some obvious connections between the "natural" world and political and social issues.

I would like to thank the *LA Times* for giving me the perch from which to carry on this dialogue, my Harcourt editor Jane Isay for finding the order in the chaos, the faculty and students of UCLA's GE70 cluster for a steady source of inspiration, and most of all GE70 alum Jenny Lauren Lee for stitching all the disparate pieces of this collection together. I'd also like to thank Roald Hoffmann and Robin Hirsch for helping include New Yorkers in this conversation through our first-Sunday-of-the-month science events at the Cornelia Street Café, Dennis Overbye for coming up with the tagline "Mind Over Matter," and Mary Lou Weisman for wisdom in all things. As the last essay ("Oops") attests, making mistakes goes with this territory. Still, my deep thanks goes out to those who have helped keep the "oops" factor in this book at a minimum, especially Janet Conrad, Mark Morris, and Tom Siegfried.

PART I

The I of the Beholder

The Emperor and Enron

You can fool all of the people some of the time and some of the people all of the time, but the easiest person to fool is yourself. Especially when the products of your own wishful thinking are also being peddled by higher authorities.

So it struck me as particularly apt that I took a class of UCLA students to the oddball Museum of Jurassic Technology in Culver City during a week when the Enron mirage was dissolving; when dubious claims for the production of fusion energy got major play in the journal *Science*; when Harvard biologist E. O. Wilson was trying to turn people's attention to wholesale extinction of life; when military planners were blithely bringing back nuclear weapons as instruments of foreign policy.

The struggle of science has always been somehow to get outside ourselves, so we can see the world objectively. The struggle has always been doomed. "We each live our mental life in a prison-house from which there is no escape," wrote the British physicist James Jeans. "It is our body; and its only communication with the outer world is through our sense organs—eyes, ears, etc. These form windows through which we can look out on to the outer world and acquire knowledge of it."

The windows are cloudy, of course, veiled by expectations, distorted by frames of reference, disturbed by our very attempts to look. Especially when we stand so close that we can't see through the fog of our own hot breath, our own smudgy fingerprints.

This is the sort of thinking bound to trail you like a wake out of the Museum of Jurassic Technology. If you haven't been there, I'm not going to recommend it. It will make you laugh, but it will also upset you. It will leave you wondering if you just didn't get it, or if you got it too well, or if someone was pulling your leg, or if you were pulling your own. You will wonder what thoughts are yours, and which are planted, and why we are so exceedingly well wired to believe official pronouncements—especially when they are obscure, pompous, and make us feel a little stupid.

Probably just as artist/creator David Wilson intended.*

For example, this funky piece of performance art might bring to mind—as it did for one of my students—Enron. How could so many people, so many accountants and investors and regulators and business journalists, believe so completely in so much thin air? At least in part because the "authorities" convinced us it was okay to do so and, worse, convinced each other.

The mind creates reality as well as muddles it. That is how placebos work.

As a friend likes to say, far more insidious than an emperor without clothes are clothes without an emperor. Authorities should always be stripped.

Scientists have thin patience for mere veneers, which is why many physicists complained loudly when the prestigious journal *Science* prominently touted a breakthrough in tabletop fusion technology—despite the widespread skepticism of the scientific community, despite failure to duplicate the results. Not that there was no there there. Just maybe there was. But tentative truths do not deserve such royal window dressing. The au-

*Wilson won a MacArthur "genius" award shortly after this column was written.

thority of science (or *Science*) is too powerful to toss around lightly.

Of course, as Jeans pointed out, our view is always partly cloudy. The closer we are, the harder it is to see. The greatest danger is believing we can ever completely separate ourselves from our surroundings. Consider, as E. O. Wilson does, the blatant absurdity of talking about the environment as if it were something apart from ourselves, a special-interest lobby, remote and foreign, like "outer space" or Afghanistan.

Increasingly, we *are* the environment.

As E. O. Wilson points out in *The Future of Life,* when humanity passed the six-billion mark, "we had already exceeded by as much as a hundred times the biomass of any large animal species that ever existed on land." We consume and exhale stuff in such huge quantities that we have already changed the air, the water, the continents. By the end of this century, we may well have extinguished half the species of plants and animals that ever lived.

We take comfort in the thought that extinction happens only to exotic creatures—big scary dinosaurs and tiny insignificant fish. Not in our backyards. The truth is, we are in serious danger of extinguishing almost everything, not excluding ourselves. And we can't stop if we can't see.

Self-referential systems are a bear. This sentence is false. Or not.

The authorities aren't helping. Instead, some authorities are telling us that we should be ready to use nuclear weapons— which do not merely destroy cities but vaporize them—to make the world safe from terror.

Somebody please get the Windex.

Normalization

I didn't understand what a truly bizarre place L.A. was until a few months after I arrived and was invited to a Halloween party. Despite considerable effort, I couldn't find an appropriate "costume": No matter how outrageous a getup I picked, I was informed it was "normal" attire for someone.

A scant year later, a friend visiting from the East Coast kept surprising me by exclaiming every time we passed a thong bikini or a crown of spiked hair.

For her, these oddities demanded attention; for me, they were already wallpaper.

So it goes. What at first glance sets off sirens in your cerebellum, after a while fails to stir up so much as a neuronal breeze.

I was horrified to see my daughter go to school with bra straps showing, only to realize that all her friends were wearing more or less the same attire.

These days, I don't even notice errant lingerie—on myself or anyone else.

We get used to almost anything that comes upon us slowly. The classic example is the frog plopped into a pot of hot water; he promptly leaps out. Place the same frog into cool water and turn the heat up slowly, however, and he'll sit contentedly until he's cooked.*

---

*I don't know whether this oft-used example is based in fact, or merely apocryphal. I've heard it argued both ways by trustworthy sources.

It can be useful not to notice things, of course. You'd be endlessly distracted if you couldn't shut out constant signals such as the feel of clothes on your skin, the glasses on your nose, the nose on your face. (And when would college students ever sleep if they couldn't tune out professors during lectures?)

Then again, some people become so accustomed to their own bad smells or foul manners that their "normal" becomes unbearable for anyone else.

Resetting "normal" means, in effect, resetting the zero point for sensation. (Physicists even use a version of this—appropriately enough called renormalization—to set unwanted effects to zero.) To register, a signal needs to rise above the background—like a car radio in a convertible. Like the stars over city lights.

It's a now classic L.A. story: After the Northridge earthquake, when L.A. went suddenly dark, hundreds of worried people called Griffith Observatory wondering about those strange lights overhead. So steadily that hardly anyone noticed, we'd been spilling city light into the sky, washing out the stars; and while a nearly starless sky seemed "normal," the sight of thousands of stars was shocking.

Leave it to the Czech Republic (a nation run by a playwright) to become the first nation to pass a law prohibiting light pollution.

Here's a scarier example: Not so long ago, the deadly microorganism known to biologists as *Clostridium botulinum* was known primarily as the source of a lethal poison found lurking in bulging soup cans—one of the most poisonous substances known. Today, it's a beauty treatment, injected at some expense into foreheads to make wrinkles go (temporarily) away. This has gotten to be so "normal" that Botox parties are today what Tupperware parties were to my mother's generation.

This is not a good thing, to put it mildly. According to the editor of *Science* magazine, Donald Kennedy, in a recent issue of his journal, *C. botulinum* ranks right after anthrax on the list of biological weapons terrorists might employ. If the demand for Botox continues to soar, and longer-lasting strains hit the market, as soon seems plausible, "will we be happy to have that many of these hot bugs around?" Kennedy asks.

The key to spinning poison into beauty potion lies partly in the words: "Botox" sounds better than "botulism." It's always the way with normalization: A "daisy cutter" seems so Martha Stewart. "Taking out" is a term from musical chairs, or for taking out the garbage—certainly not somebody's son or grandmother. No splattered blood or melted flesh.

In reality, a daisy cutter is, as everyone now knows, a fifteen-thousand-pound "antipersonnel" weapon. It delivers scarcely one-thousandth the power of the Hiroshima nuclear bomb—which itself is pathetically puny by today's (even mininuke) standards.

What does it mean to scale up by a thousand? Imagine, as a physicist friend did, that you suddenly find yourself serving dinner for four thousand people instead of four. Making do with the same kitchen, same pots, same glassware. This is a fair comparison, he said, because after all, the Earth itself—the people, the homes, the civilizations—does not change even as firepower increases.

The progression from conventional weapons to nuclear ones is not like going from bra straps to tongue piercing. Closer to summing up the situation is Einstein's remark: "I do not know with what weapons World War III will be fought, but World War IV will be fought with sticks and stones." Good people will differ on choices of action, but you can't see where you're going if you mistake your destination for wallpaper. Be careful what you normalize. It might just take you out.

Blindsighted

The smoking gun. The incriminating tape. The writing on the wall.

We think we know what solid evidence is. We think we know what we see with our own two eyes.

Of course, we don't. And there's no better evidence of that than the recent spate of criminal convictions overturned on the basis of DNA analysis—convictions often based largely, sometimes solely, on eyewitness testimony.

What we see with our own two eyes is often no evidence at all.

Consider a series of experiments conducted by Dan Simons at the Harvard Visual Cognition Lab. In one, you watch several groups of students toss balls to each other and count the number of completed passes. Simple enough. Except that when you view the video a second time—this time without bothering to count the ball tosses—you see a man in a gorilla suit stroll right through the ball tossers, turn to face you, and pound his substantial hairy chest. Most people don't see the gorilla when their eyes are on the balls because of what Simons calls "inattentional blindness"—a trick long known to magicians and crooks.

In other equally creepy experiments, Simons shows how you can be merrily giving directions to a stranger—looking that person right in the eye—and not even notice when a different stranger is suddenly substituted. A person of a different height, with different hair, wearing different clothes.

So much for eyewitness testimony.

We see what we expect to see, don't see what we don't expect. In the 1930s, during the height of anti-Semitism, the presence of large numbers of Jewish players in pro basketball was interpreted by at least one sports writer as evidence of their "scheming minds," "trickiness," and "general smart-aleckness," according to Michael Shermer in the *The Borderlands of Science: Where Sense Meets Nonsense.*

On the other hand, even firm evidence has been easily dismissed when it rubbed conventional wisdom the wrong way.

For example, although anyone can see with their own two eyes that Africa and South America fit together like pieces of a puzzle, it was a long time before even scientists believed that the two had once been part of a single landmass. Until plate tectonics came along, no one imagined a mechanism that could motivate a continent to move.

In science, evidence rules. So it's not surprising that scientists have come up with a trick or two for sorting out the firm from the flimsy.

What qualifies as "firm" under their criteria might surprise you. For example, in the fall of 2001, results from the Fermi National Accelerator Laboratory suggested that barely-there particles called neutrinos behaved just a smidgen differently than theories said they should. And yet, this smidgen (a 1 percent deviation, to be exact) was greeted by scientists as a possible crack in the foundations of physics.

Why should such a tiny mismatch mean so much? In this case, the reasons have to do with both the solidity of the theory and the precision of the new results.

That isn't always the way: during the same month Brookhaven National Laboratory had to retract a similar announcement of "new physics" based on a similar smidgen of difference between theory and experiment. While the measurement stands more solid than ever, the theoretical predictions, it turned out, were thrown off by a misplaced minus sign.

Of course, the theorists were motivated to check the calculation precisely because of the mismatch in the measurement. And that's the crucial point: The mismatch itself, whether it's between theory and experiment or eyewitness testimony and DNA, is clear evidence that something is wrong with our understanding—of how the universe works or how the justice system does.

If science has taught us anything, it's that no one line of evidence is ever enough. Confused and frightened people can confess to crimes they didn't commit. DNA samples can be contaminated or even planted. Scientists make errors in calculations or let wishful thinking skew their reading of the "facts."

At the same time, when all the evidence points in the same direction, you have to follow where it leads, no matter how unsettling the result.

An obvious example is the birth of the universe from nothing and nowhere in a primordial explosion—a patently absurd idea if there ever was one. Yet it's widely accepted because so many different lines of evidence point clearly in that direction: the motions of galaxies, the shape of the universe at large, the composition of matter.

It's strange but true: You can be more certain of what happened 13 billion years ago when no one was around to see it than you can of what happens today before your eyes.

When it comes to evidence, eyewitness testimony can't hold a candle to the Big Bang.

Uncertainty

Everyone understands uncertainty. Or thinks he does."
So says German physicist Werner Heisenberg in Michael
Frayn's play *Copenhagen*.

I hate to argue with someone no longer around to defend
himself, but I beg to differ. I think most people don't begin to
understand uncertainty, which is why we run from it screaming.
What we want, and believe we can have, are guarantees: Is air
travel safe? Does eating broccoli prevent cancer? Is the global
climate warming?

He loves me; he loves me not.

Plucking petals off daisies or reading the stock market
ticker, we all want to know for sure.

Alas, our universe is an uncertain sort of place, as Heisenberg discovered. Every bit of certain knowledge comes at the
price of uncertainty someplace else.

Take, as Heisenberg did, an elementary particle. Try to pin it
down: Where, exactly, is it? To know, you must measure it, and
to measure it, you must disturb it in some way. Thus, the
minute you pin down its position, you no longer know exactly
how it is moving. Conversely, the more precisely you pin down
its motion, the less clearly you can see its position.

This apparent limitation to human measurement turns out
to be a fundamental property of the universe. Even the vacuum
of empty space quivers with uncertainty; in fact, these incessant
nervous twitches in nothing may well add up to more energy
than everything else in the universe combined.

Frayn's play, likewise, takes place in a cloud of unanswered questions.

Why did Heisenberg go to Copenhagen in 1941 to see his old mentor, Danish physicist Niels Bohr? At the time, Heisenberg was the head of the German nuclear weapons effort. Did he plan to pry atomic secrets from his friend? Or was he trying to enlist Bohr's help to try to prevent atomic bombs from being built?

And why did Germany fail to build an atomic bomb? Was it because Heisenberg deliberately withheld information from the Nazis? Or because this consummate mathematician neglected to do an obvious calculation?

Did Heisenberg know himself? Is there a single answer?

Real people, like particles, are complicated. Their states are entangled—with each other, with forces, with history. Physics is entangled with politics and the personal gets entangled with everything. Bohr was entangled with Heisenberg, and so was his physics.

Though Bohr is the acknowledged "father" of quantum mechanics, perhaps his best-known contribution is his theory of complementarity—a set of ideas valued as much by philosophers as by physicists. Position and velocity comprise a complementary pair; each is inherently uncertain. Only together can they form a complete picture of reality.

At the same time, complementary descriptions of nature are mutually exclusive. To predict where a particle will go, you need to know both its position and velocity, but measuring one precludes knowing the other. Measured one way, an electron is a particle; measured another, it is a wave. To see the wave aspect, you destroy the particle; to see the particle, you destroy the wave.

Nature is full of such dualities. A human being has infinite worth. A human being is merely a collection of electrons and quarks. Music is a mathematical relationship among acoustic tones. Music is a means of emotional expression. Both descriptions are true, and neither is complete.

Bohr and Heisenberg, too, made a complementary pair. Bohr relished paradox. Heisenberg was undone by it.

Bohr wasn't content unless physics made palpable sense: "What does the mathematics *mean* in plain language?" asks Bohr in the play. "What are the philosophical implications?"

Heisenberg found truth in equations: "What something means is what it means in mathematics. Mathematics *is* sense!"

(Yet even Heisenberg finds room for doubt: "Mathematics becomes very odd when you apply it to people. One plus one can add up to so many different sums.")

The physicists were old friends, like father and son. They were also enemies: Germany occupied Denmark.

Together, they created a physics of deep truth and beauty— physics that gave people (as observers) a central role. Together, they also made possible physics with the power to destroy all people once and for all.

If that were to happen, says Bohr's wife, Margrethe, in the play, "even the questions that haunt us will at last be extinguished."

As long as there are people, of course, there will continue to be questions—and the hope for certain answers. And there lies a paradox Bohr would have loved: Faith can be certain, but knowledge is always to some extent elusive. Some answers are like particles—everywhere, nowhere.

Different depending on how you ask, or when.

The price of knowledge is not sin but uncertainty. Easier to swallow, perhaps, but far more difficult to comprehend.

Murmurs

When the universe speaks, astronomers listen.
When it sings, they swoon.

That's roughly what happened recently when a group of astronomers published the most detailed analysis yet of the cosmos's primordial song: a low hum, deep in its throat, that preceded both atoms and stars.

It is a simple sound, like the mantra "Om." But hidden within its harmonics are details of the universe's shape, composition, and birth. So rich are these details that within hours of the paper's publication, new interpretations of the data had already appeared on the Los Alamos web server for new astrophysical papers.

"It's stirred up a hornet's nest of interest," said UCLA astronomer Ned Wright, who gave a talk to his colleagues on the paper—as did so many others—the very next week.

So what is all the fuss about? Why are astronomers churning out paper after paper on what looks to a layperson like a puzzling set of wiggly peaks—graphic depictions of the sound, based on hours of mathematical analysis?

Because there's scientific gold in them there sinusoidal hills.

The peaks and valleys paint a visual picture of the sound the newborn universe made when it was still wet behind the ears, a mere 300,000 years after its birth in a big bang. Nothing existed but pure light, sprinkled with a smattering of subatomic particles.

Nothing happened, either, except that this light and matter fluid, as physicists call it, sloshed in and out of gravity wells, compressing the liquid in some places and spreading it out in others. Like banging on the head of a drum, the compression of the "liquid light" as it fell into gravity wells set up the "sound waves" that cosmologist Charles Lineweaver has called "the oldest music in the universe."

Then, suddenly, the sound fell silent. The universe had gotten cold enough that the particles, in effect, congealed, like salad dressing left in the fridge; the light separated and escaped, like the oil on top.

The rest is the history of the universe: The particles joined each other to form atoms, stars, and everything else, including people.

"The universe was very simple back then," said Caltech's Andrew Lange in one talk. "After that, we have atoms, chemistry, economics. Things go downhill very quickly."

As for the light, or radiation, it still pervades all space. In fact, it's part of the familiar "snow" that sometimes shows up on broadcast TV. But it's more than just noise: When the particles congealed, they left an imprint on the light.

Like children going after cookies, the patterns of sloshing particles left their sticky fingerprints all over the sky.

The pattern of the sloshes tells you all you need to know about the very early universe: its shape, how much was made of matter, how much of something else.

The principle is familiar: Your child's voice sounds like no one else's because the resonant cavities within her throat create a unique voiceprint. The large, heavy wood of the cello creates a mellower sound than the high-strung violin. Just so, the sounds coming from the early universe depend directly on the density of matter, and the shape of the cosmos itself.

Astronomers can't hear the sounds, of course. But they can read them on the walls of the universe like notes on a page. Compressed sound gets hot and produces hot splotches, like a pressure cooker. Expanded areas cool. Analyze the hot and cold patches and you get a picture of the sound: exactly how much falls on middle C, or B-flat.

What they've seen so far is both exciting and troubling. The sound suggests that the universe is a tad too heavy with ordinary matter to agree with standard cosmological theories; it resonates more like an oboe than a flute. Something's going on that can't be explained. The answers will come as even more sensitive cosmic stethoscopes listen in over the next few years.

Lest you think these sounds are music only for astronomers' ears, consider: The same wrinkles in space that created the gravity wells that gave rise to the sounds also blew up to form clusters, galaxies, stars, planets, us.

Even Hare Krishnas murmuring *Om*.

Eclipse

S till, it moves."
    So, it is said, Galileo muttered under his breath in the seventeenth century after publicly recanting (under threat of torture) his heretical revelation that the Earth moves around the sun.

Today we think we know the Earth moves. We think we know it's a tiny blue marble floating in the black vastness of space. We think we know that our grasp on solid ground is only a highly local effect of that thin glue we call gravity.

But we don't, not really. For all our scientific knowledge, we believe in our hearts that we walk around on a steady Earth under the benign influence of an endless sky, that the eternal sun provides life support just for us.

Nothing shatters this smug self-assurance like a total solar eclipse—that eerie darkness at noon that falls when the moon places itself in the path of the sun's light, sending a palpable chill through the atmosphere.

No wonder people become unglued. When the moon throws its cold shadow on our world, it cuts our connection with the star of the solar system. It exposes our fragility for all to see. It breaks the spell, dispels the illusion. Like turning up the houselights during a movie.

At once, we see ourselves for what we are. It's a close encounter that pulls viscerally on our souls as surely as gravity pulls the Earth into the sun's embrace.

Even astronomers—who study this stuff—can't account for its power.

An eclipse results from a choreography of coincidence that allows the tiny moon to cover the enormous sun—93 million miles away, and a million times larger than the Earth. The diameter of the sun is four hundred times that of the moon, but the sun is also four hundred times farther away. Thus the moon can completely blot the sun just as your thumb held at arm's length can completely cover a far-off mountain. In a matchup made in heaven, the tiny lunar interloper completely snuffs out the star of the solar system.

But like most life-shattering experiences, an eclipse sneaks up on you.

So it was that I found myself scanning a brilliant blue sky over the Black Sea some summers ago, finding warning of what was to come. Only in your imagination could you see that the moon was slowly making its way toward the momentous encounter. Only in your mind could you perceive the slow dance of the solar system that whirls the moon around the Earth and the Earth around the sun in wide elliptical arcs.

Even when the moon took its first tenuous bite out of the sun, it did so unseen, slipping in under cover of the sun's glare. A stealth moon.

Through dark protective glasses, you could watch the size of the bite grow, the gouge increase. Without them, you noticed nothing special until about fifteen minutes before the blackness set in.

Then, suddenly, shadows turned razor sharp. Fingers turned into claws, and the hair standing up on arms turned people to beasts. The light grew strange, unnatural.

And then, the moon hit the spot. Shades of night inexplicably dropped during the middle of the day, chilling the air like a ghost. Everything was upside down. There was a glow around the horizon, but the sun was setting "up there."

Except there was no such thing as "up there" anymore. We were one of three spherical worlds slow-dancing in the solar system. More than one: As the moon clicked into place, Venus lit up on the left; Mercury blinked on from the wings at far right.

Blacker even than space, the disk of the sun sprouted a fringe of pink lace—loops of electrified gases leaping out from the star's surface. Long gossamer threads of glowing "hair" spread far out in all directions.

Then, bingo! A flash of light, and the sun became a diamond ring out of a Tiffany advertisement—a golden circle with a spot of brilliant fire where the last rays leaked out between some high mountains on the moon.

At last, reluctantly, the moon let go, and the three worlds once again went their separate ways.

But the chill remains.

Like most shadows, the shade of the moon is scary, and the eclipsed sun throws an unwelcome light on the precariousness of planetary existence. The universe is cold, dark, vast, empty. The sky is only skin-deep.

The otherworldly quality of eclipses comes at least in part from the inescapable revelation that we are not the world, but only one of many worlds.

It moves. They all do.

Humility

The entire history of science, it often seems, is one long lesson in humility.

From what we once thought was our privileged spot at the center of the cosmos, we fell into the arms of a very ordinary spiral galaxy, one of billions in the universe. Our solar system, astronomers tell us, congealed out of the debris of long-dead stars the way fat congeals in soup.

To rub it in, cosmologists now tell us that even the universe is probably nothing special—just one of a very large litter of universes that may well breed like bunnies.

It's even possible, some physicists say, that we experience only a small sliver of the true dimensions of space—confined like pond scum to the surface of a much deeper universe whose depths we will never fathom.

Physicists and astronomers are mostly to blame for this humiliating state of affairs.

But biologists have certainly done their share. Is it any wonder that the Kansas state school board couldn't stomach Darwin's lesson that we're all descended from hairy, small-brained hominids? Teenagers aren't the only ones embarrassed by strange-looking parents.

And as the late biologist Lewis Thomas put it, the situation is far worse even than that. We can now trace our ancestors back to bacteria: "We go back to *it,* of all things," he wrote in the foreword to biologist Lynn Margulis's delightful book *Microcosmos.*

Microbes, Margulis writes, not only rule the earth, they rule us as well. Fully 10 percent of our dry body weight consists of bacteria. We depend on them for digestion, among other things. In a very real sense, they invented us for their own purposes, as warm salty environments in which to breed and prosper.

Life has been around in simple form for more than 3.5 billion years; the entire history of humanity accounts for less than 1 percent of that. Even now, we don't amount to much. There are nearly a million species of insects, compared with a paltry couple of thousand species of mammals.

Lest you think we've become somehow separate, superior, to all of this, consider: We still share 50 percent of our genes with fungi. In early gestation, a human embryo is all but impossible to distinguish from that of a pig, ox, or rabbit.

And what about plants? Without them, we would not only go hungry, we wouldn't be able to breathe. If all photosynthesis ceased, the oxygen in Earth's atmosphere would be mostly gone in less than a million years.

Even dumb rocks show us up, rattling and rolling us out of our beds at the slightest provocation—whenever one continental plate rubs its neighbor the wrong way.

Chemists, in the last analysis, are probably the worst. Lust, fear, hate, even spiritual enlightenment—all our emotions boil down to disembodied dances with molecules. Our most profound thoughts are nothing but neurotransmitters blinking messages to each other inside the blackness of our brains.

In fact, cognitive scientists, following Freud, tell us that only a very small percentage of our thoughts even surface to the level of consciousness—meaning that most of the time, we don't even know what we think.

It's almost as if nature has set out to reward those who would learn her secrets with a single, all-encompassing lesson: You don't amount to a hill of beans.

But there's another way of looking at this. In some sense, we are more central, more powerful, than ever.

When people were seen as sitting at the center of the cosmos, they also were at the bottom of the heap. Human beings were the powerless and imperfect playthings of all-knowing, all-powerful gods who controlled everything from plague and pestilence to the course of weather and love affairs. The heavens revolved above and beyond on precision-crafted crystal spheres, forever beyond our sullied reach.

Today, we know we are formed from the same stuff as stars. We are a piece—not separate—from this grand, entangled cosmos. And we control—at least to some extent—our destinies.

More impressive still, our brains have learned to spin spacetime into black holes and to tame quantum unruliness to operate computers. On good days, we even have the mind to appreciate the grandeur of it all—setting it to music, or art, or equations.

Along the way, of course, we've also learned enough to destroy much of it without the help of any gods at all. This, in itself, is an awesome sort of power.

So it's not entirely clear whether the discoveries of science have really made us more humble. But they have certainly, as the late physicist Frank Oppenheimer put it, "changed the nature of our humility."

Inside Out

An artist friend gets a huge kick out of people who harbor romantic dreams of traveling to "outer space." After all, he points out, we already live in outer space. Our little globe floats around in vast blackness, barely making a dent in the fabric of space-time. Space travel is an everyday event. In one year, we go all the way around the sun; in a few hundred million, we'll hitch a ride with our solar system around the entire Milky Way.

Scientists and science writers fall into the same fallacy when they speak of entities like "dark-matter" particles as if they only existed "out there," in "space."

But as Case Western Reserve University physicist Lawrence Krauss reminds us, "These exotic dark-matter particles [are] 'in here' as well as 'out there,' traversing the Earth, terrestrial laboratories, and even ourselves."

The confusion between inside and outside pervades our thinking about science.

Consider: Every quality we ascribe to the things in the outer world flows from properties of subatomic particles. Color, flavor, smell, texture, hardness, electrical conductivity—all depend on the invisible architecture of atoms. In this sense, inside *is* outside.

Making false distinctions between the insides and outsides spreads misleading ideas about science to students as well. I once heard a chemistry teacher tell a group of high school students they wouldn't like chemistry because it was so "abstract"—so far outside the average teenager's realm of experience.

On the contrary, chemistry is the force behind just about everything in a teenager's life: love and sex, fear and loathing, pizza and fries, cosmetics and cars. There's no way to separate the sensations on the outside from the chemistry going on underneath.

In the natural world, inside and outside tend to merge with no clear wall. The outer membrane of a cell, like the atmosphere of Earth, is permeable. Our skin keeps out wind and rain, but it can't keep out most radiation and many kinds of matter. As you read this, signals from radio and television stations are zooming right through your body—along with stray cosmic rays and countless neutrinos.

The ozone layer keeps out most ultraviolet radiation, and the Earth's magnetic field diverts most of the solar wind toward the poles, but often, some of both seep through. UV rays cause skin cancer, and electrically charged particles blown off the sun disrupt earthly radio transmissions. Recently, a blast from a "magnetar" halfway across the Milky Way traveled more than 23,000 light-years and rocked Earth's ionosphere. The fact that it reached out to touch us from so far out in space, one scientist said, reminds us that Earth does not live "in splendid isolation."

Indeed, one of the great turning points in physics was Isaac Newton's realization that heaven and earth did not operate on different sets of natural laws. Before Newton, people believed that apples fell because of gravity, but planets were pushed around by angels.

Newton showed that there was no real difference, physically speaking, between the forces that rule the heavens and the

Earth. The same gravity that pulled the Tower of Pisa askew keeps the moon in its orbit.

Ever since Newton, scientists keep finding more common ground between Earth and outer space. They've found volcanoes on Jupiter's fiery moon Io, and signs of a vast ocean under the ice on Io's sister, Europa.

Pathfinder's sojourn on Mars showed that floods and sandstorms and dust devils are, or have been, as common there as on Earth. Perhaps similar life-forms existed in wetter, warmer days of the red planet's evolution.

In the end, nothing turns logic inside out so much as contemplating the shape of the universe itself. Is it infinite, or does it end? And if we do live in a cosmos with an edge, what's on the outside? Is it different from what is "inside"?

As it turns out, the universe can be finite, yet endless. Think of the surface of Earth, which has a definite surface area, but no "edge." You can start walking north, and continue round the globe until you wind up where you started. If the universe has a similar shape, you could theoretically shoot a rocket into outer space and it would eventually come back and hit you in the head.

What begins as inside emerges from outside. And vice versa.

Weird Science

Nice try, Hollywood.

I don't care much for scary movies, but I just couldn't resist an invitation to see the newly released version of *The Exorcist*. So I prepared myself to be scared silly by swiveling heads and earthshaking devils and levitating beds.

And guess what? The scariest thing in the movie is the very realistic depiction of arrogant, know-it-all doctors torturing their young patient with modern medical technology (and oh, the blood!!) while in the end offering nothing better than a prescription for Ritalin. Now *that's* scary.

As usual, real life trumps fiction. In spades.

No matter what bizarre scenarios Hollywood dreams up, Nature has done it before, and better. Even the most imaginative moviemakers can't come close to the terrors and wonders of the real thing.

I mean, take your swiveling heads and levitating bodies—or even Linda Blair's newly inserted spider walk down the stairs. What is that compared to, say, leprosy? The "Elephant Man" disease? Or plague?

Or how's this for a scenario? Virus in African monkeys gets transmitted across species to infect humans on a global scale—wiping out huge segments of the population in some countries? Or how about flesh-eating bacteria? Or human-concocted terrors like genital mutilation?

Not to mention the everyday horrors like the millions of dust mites that share your bed every night; the microscopic monsters that live in your eyelashes.

Of course, horror is only one genre where Hollywood takes second place. When it comes to good old-fashioned gee-whiz special effects, it can't hold a candle to what's really out there: mysterious bursting objects in space that can spit more light into space (at least for an instant) than entire galaxies. Black holes that gobble stars whole and bring time to a halt.

The Matrix was fun, but at most, its characters were oozing in and out of four dimensions of space. This is child's play for today's physicists, who routinely explore the tangled topologies of ten or eleven. Not to mention two or more of time.

Our entire universe, according to these physicists, may be like the scum on the surface of milk: a membrane floating on a much larger cosmos of many more dimensions. Like the bad guys in *Superman,* we would be trapped inside a thin slice, unable to communicate with the "real" world outside.

And then there's the Alice-in-Wonderland world of particle physics, the lilliputian realm so spooky that even Albert Einstein refused to believe it was real.

Particles communicate like ghosts—telepathically, you might say—across vast distances, even without contact. Particles can be here and there, exist and not exist simultaneously.

Schrödinger's famous cat—the one that's both alive and dead at the same time—is no longer a fable; physicists have actually measured particles in the lab in a similarly schizophrenic state.

And don't even get me talking about cosmology. At a typical cosmology meeting these days, you're likely to hear tales of universes that breed like rabbits out of hats and pop out of nothing at all for no apparent reason, even evolving like species: According to some theories, the flip side of a black hole is a so-called white hole—which is indistinguishable from the "big bang" that created our universe. Every time a star collapses into such an abyss, a new universe appears.

In the ultimate scheme of things, we may all be the progeny of black holes.

And let's not leave out technology. How about micromachines that can crawl inside your body to diagnose disease, or deliver medicines? Machines that can just about assemble themselves? Particle accelerators that re-create the birth of the universe (if on a small scale)—thousands of times a second?

Or math, for that matter? What's a Hollywood ghost compared to an imaginary, surreal, or transcendental number?

I could go on (and on). T. Rex was dreamed up by evolution long before Steven Spielberg came along, as were carnivorous plants.

However you cut it, Hollywood isn't half as creative as good old Mother Nature. Science is stranger than fiction.

In the end, reality always writes the better script.

Love and Bosons

L ove is a boson. Who knew?

This insight came from a physicist friend during a conversation reflecting on how some scientists never seem to get enough approval. No matter how many prestigious awards they win, no matter how big their reputations, they remain somehow insecure.

And, my physicist friend concluded, "That's because fame, like love, is a boson."

Now *boson* may not be a household word, but it does account for one of two major families of particles that make up our universe.

A particle of light, for example, is a boson.

My friend said love is a boson because bosons don't know the meaning of "enough." Bosons can clump together without limit; squeezing inside the smallest space, there is always room for more. Bosons are bottomless pits.

Contrast this gregarious behavior with that of fermions, the other main branch of the particle family tree. Notoriously standoffish, fermions are loners; only one can occupy the same place at the same time.

This explains why, for example, you can rest your head on the table without falling right through. The table is made of atoms, which in turn are mostly made of empty space. That space is very lightly populated with fermions, which are like those people who sit in airplane seats with their arms and legs spread out, taking up everyone else's room. Even though they're small, they manage to preclude anyone else from invading their territory.

By contrast, you can put your head through a beam of light because light is made of bosons.

Bosons and fermions appear to be completely unrelated. But physicists believe that long ago in the very early universe they were very much the same. During this brief honeymoon of primordial harmony, bosons and fermions were all related to each other in one big happy family—something like the Nelsons.

Today, the family is so badly broken up that the two sides seem to have nothing to say to each other—more like the Capulets and the Montagues. Physicists believe that if they could understand what split up the family in the first place, they could explain why and how bosons and fermions and everything else in the universe came to be.

What's more, many physicists believe that each fermion has a long-lost bosonic alter ego hidden in the family closet, and vice versa. They believe that if they put enough energy into their particle accelerators, they could make these long-lost significant others reappear.

It may seem odd to talk about particles in terms of love and family, but in fact, all of particle physics is a study of relationships. What interacts with which? Who is related to whom? How often does so-and-so interact with such-and-such?

Charting family genealogies is big business for particle physicists. Like species, particles mate and beget progeny according to very specific sets of rules, often obscure to outsiders. Only certain pairings are possible, under certain conditions, in certain prescribed ways.

Supposedly, the family genealogies are clear and you always know who is who. In reality, it's a lot more complicated. For example, in the exotic state of matter known as superconductivity, supercold fermions clump together just like bosons, overcoming their innate reclusiveness to become one with their fellows.

What's more, there is a theoretical way to create protons and neutrons—the fermionic constituents of the atomic nucleus—out of bosons. This idea was around in the 1950s but ignored until recently, when physicist Edward Witten "made it respectable," according to MIT's Frank Wilczek.

Wilczek is the person other physicists turn to when they have relationship problems with bosons and fermions. (Call him Dr. Frank.) He dreamed up a particle called the "anyon," which mediates between the two. In three dimensions, particles can only be bosons or fermions, according to Wilczek. But in two dimensions, they can be any combination of the two: thus, the anyon. To everyone's surprise, his theoretical anyon turned out to be real; it shows up in a highly esoteric phenomenon of two-dimensional surfaces called the quantum Hall effect.

Particles and families are alike in other ways as well. Certain relationships are doomed, for example, and everyone knows it. They only don't know how the breakup will occur, or when. Gravity, for example, appears to have irreconcilable differences with quantum mechanics.

Particle families also have their odd ducks (there's one called the strange quark, for example) and missing relations (like magnetic monopoles, which should be out there somewhere but have never been seen). Newcomers are not always welcomed into the fold: "If I could remember the names of all these particles. I would have been a botanist," physicist Enrico Fermi grumbled during the 1950s when new kinds of particles seemed to be appearing at the physicists' doorstep almost daily.

But mostly, particle families are like human families because despite the most dramatic differences, you frequently find deep connections. Just as TV's inimitable *Sopranos* reminds us that the family of a New Jersey mob boss has the elements of all families everywhere.

The more we get to know particles and people, the more we learn about intricate past histories, hidden entanglements, unexpected complications. What seems understood and simple in one moment can change in an instant into something strange and complex.

Love is like that. So are bosons.

Seeing

A woman I know was told by three different doctors that the large irregular blob on her mammogram was serious cause for concern. From the looks of it, all three said, it appeared to be a malignant tumor.

But as any scientist can tell you, "from the looks of it" is a phrase that's fraught with peril. Happily, a fourth doctor—more experienced in reading mammograms—decided to take some more pictures from different angles, and lo and behold, the tumor was only an illusion.

The art of seeing is more subtle than most people think—even in such seemingly straightforward fields as physics and astronomy. Yesterday's newfound planet is revealed as a wiggle in the instrument that took its picture; that newly discovered subatomic particle turns out to be a stray piece of background noise.

It happens the other way, too—when the background noise turns out to be a new particle, or the wiggle that was dismissed as an aberration turns out to be a planet.

Scientists, like doctors, spend most of their time studying things they can never directly see. This means they have to become experts in the art of seeing the obscure, and sometimes the invisible.

Take dark matter—the unseeable stuff that supposedly makes up 90 percent of the universe. To catch it in the act of holding the universe together, experimenters must devise all manner of clever traps—except they don't know what kind of bait to use because they can't agree on what they're looking for.

Or take Earth, which is right under our noses. The most interesting slice of it may well be three days' drive straight down—where the outer core meets the molten mantle. And there's no way to get there from here.

Instead, geophysicists, like doctors, look for signs on the surface that tell tales of hidden goings-on underneath—say, scars on the planet that tell of ancient upheavals. Or they resort to reading images created by sound waves reflected from structures inside the earth. These sonograms are similar to the ones physicians use to see inside the human body. An obstetrician getting a look at a developing baby bounces sound waves off the fetus to get an image; a geophysicist rattles the ground with explosions to set off sonic booms inside the earth. Astronomers even use a version of sonograms to study the sun and stars by watching the way they ring like bells under the influence of internal vibrations.

When just looking isn't enough, scientists may resort to a version of surgery—cutting holes in the planet's skin to bring up samples to be studied in the lab. Or they try to re-create events they can't see—like growing cells in a petri dish, or squishing materials together under enormous pressure to simulate, say, the inside of the giant planet Jupiter.

Paleontologists must find ways to peer into the past, re-creating the detailed physiology and family habits of dinosaurs from a few fossilized footprints, a jawbone, some dusty DNA. Climatologists must look both backward and forward in time, using powerful computers to turn eons into seconds, then rewind, throw in some new theories, and try again.

Chemists have come up with myriad ways of picturing the invisible molecules that are the tools of their trade. Cornell chemist Roald Hoffmann compares a half dozen images of the canphor molecule as it appears in different familiar representations: a chemical formula, a line drawing, a ball and stick model, a computer simulation.

Which does the molecule really look like? he asks. Although all four images depict it accurately, the molecule "looks" like none of them.

Some scientists are forced to follow clues that seem far removed from their subject matter—looking for the origin of human life in dust grains exhaled by passing comets, for example. Others fiddle with equations to tease out patterns that may reveal unseen forces. Still others use statistics and clever mathematics to find needles in haystacks, or tell whether the smudge in the telescope is a star or galaxy.

And nothing trains a seer like a lifetime of looking. Astronomer Vera Rubin, studying a galaxy that had been seen and overlooked by many astronomers before, saw the first-ever case of rotation in what appeared to be the "wrong" direction. She did it by taking two years to "make friends," as she put it, with this unusual cluster of stars.

My personal favorite is the story of antimatter. A theorist discovered it as a minus sign in an equation; an experimentalist saw it as a cosmic ray track that curved the wrong way. What's remarkable is that neither gave in to the temptation to dismiss such an unexpected discovery as a mere aberration, or noise.

Today, of course, antimatter is used routinely by doctors to see inside the human body (in the form of PET scans, or positron-emission tomography).

When all is said and done, being a good looker may be the most important quality a scientist—or a doctor—can possess.

Moving Mountains

Sometimes, you can't *not* take it with you.

I was reminded of this one summer during a week stalking wildflowers and crunching through glaciers in the wilderness of the Canadian Rockies.

A friend stopped to marvel at the miniature tableau created by the chance alignment of twig, brook, flower, mushroom, moss.

I stopped to marvel at the way gravity had bent a sapling backward into a parabola.

"Just what you *would* notice," she said, rolling her eyes.

Even this far from civilization, I couldn't help packing a certain attitude—a way of seeing things through the eyes of science. Here, in what many experienced hikers consider to be the most beautiful spot in the world, a realm of pure spiritual transcendence, I couldn't help seeing geology, physics, biology, psychology, and, yes, even math.

To be sure, mountains can move us. But it's easy to forget that mountains also move, capturing as they do the history of the planet and everything that ever happened on it or to it. Traveling in the mountains is a trek through petrified spacetime.

The Canadian Rockies, in particular, have the look of a construction site: colossal fossilized ocean waves, frozen while breaking, fierce as sharks' teeth, menacing as shards of glass.

Underneath, the earth is doing a slow boil. "The Earth's mountain ranges have been built by nuclear energy," Ben Gadd reminds us in *The Handbook of the Canadian Rockies.*

The heat released by radioactive decay melts rocks, driving currents that push the continents around. When they plow together, mountains rise up in smooth folds, like a gathered tablecloth, exposing the layers that were laid down eons ago in the ocean—the sands of time.

The sharp edges come later, with the ice. "I thought glaciers were supposed to move," said my friend, impatiently watching the fields of snow for signs. "They do," said I. "Only glacially."

Water is strong stuff. It can chisel stone.

It also does pretty things with optics.

A raindrop hanging from a pine needle forms an upside-down image of the mountains, the clouds, the lake—a world inside a drop of water.

In the lakes, the color of the water turns intense, gaudy, oddly unnatural—a turquoise Barbie might wear. The rich hue is the result of the uniformity in size of the dust particles— "rock flour"—washed off the mountains by the glaciers. The heavier particles sink to the bottom of the still, cold lakes, leaving only the smallest suspended. Like molecules in the air that selectively scatter the blue light of the sun to our eyes, the particles in the lake scatter only aqua green.

The lakes mirror everything, multiplying limbs of long-dead trees—picked clean by the elements—into skeletons of bizarre creatures. These illusionary life-forms look uncannily like the real fossilized remains found in the famous repository known as the Burgess Shale, not far away.

Captured in the shale is one of the most prolific outpourings of creation the world has ever seen—creatures long extinct, some with five waggling eyes or seven spiky legs.

But then, all life seems bizarre up here in the thin air. Trees grow tall and spindly, gasping for breath. They poise precariously on cliffs, bend like elbows, hang upside down, holding on by their roots—a lesson in tenacity.

Miniature mop-headed plant life springs up absurdly by the streams—alien species, something out of Dr. Seuss.

We nearly trip over a couple of marmots sunning on the trail. Marmots adapt to the harsh climate by sinking into something like a near-death experience for seven months of every year; their hearts scarcely beat, their lungs barely breathe.

There are critters we city people could do without. Bears, for example. A thrill to see at a distance, but a source of terror (at least for me) on the trail. At some level, we like our wilderness as innocuous as Disneyland.

Which brings us to psychology. While I worry about the grizzlies, people here shake in their boots at the thought of coming to L.A. Wild life, like so much else, is a matter of perspective.

And math. What is the probability of getting eaten by a grizzly, anyway? It's not at all like the probability of getting heads on the toss of a coin. For one thing, the probability of a coin landing heads is not affected by the outcome of previous tosses. However, the probability of running into a bear is most certainly affected by the fact that a bear was seen on this very trail only yesterday.

We discuss the mathematics of probability very LOUDLY so the bears will hear us coming.

There is neuroscience here, too—the way the mountains play on our emotions. On his first extended trip to the Sierras, mountaineer John Muir let loose such a "wild burst of ecstasy," he reports in his diary, that he frightened a bear.

Is it an accident that gods invariably occupy mountaintops?

"Why, after all, should one get excited about a mountain," asked the essayist Wendell Berry, "when one can see almost as far from the top of a building?"

Why, indeed?

The science may escape us. But the truth is as stubborn as stone: Mountains move us in a way we will never move mountains.

Uncommon Sense

If there's one quality that's sure to get a scientist into trouble, it's common sense. Over and over again in the history of science, common sense has been exposed as a lousy guide to truth.

Take astronomy, for example. In 1825, the French philosopher Auguste Comte argued that humankind could never find out what stars were made of. And what could be more sensible? Anyone can see that stars are far too remote to analyze in a lab.

Within a century, however, astronomers had learned to read the bright and dark lines in the spectra of starlight for clues not only to the physical makeup of stars, but also to their temperature, age, and motion.

Or take biology. Who would have thought that our very bodies were populated with hordes of species of bacteria before Leeuwenhoek came along and looked at saliva through his first crude microscope? Today we know that the vast majority of life-forms on earth and in the oceans become visible to us only through microscopes; we fuzzy mammals are the lumbering oddballs.

Much more recently, biologists came to the commonsense conclusion that life needed sunlight and oxygen to survive. No sooner had they concluded that than whole colonies of life-forms were discovered in the total darkness of the ocean floor, living off sulfurous fumes steaming from boiling hot ocean vents.

Indeed, the commonsense view that life requires oxygen completely distorts the historical record. Free-floating oxygen is a newcomer to our planet, a toxic poison released into the atmosphere by the first green plants.

And contrary to the conventional wisdom that pouring carbon dioxide into the air "pollutes" our planet, it actually serves to bring it back to its natural state. (Humans, of course, are not "natural" to Earth and would not survive such a dramatic change. But that is a detail.)

Even geologists, who study the comparatively sedate Earth and planets, haven't been immune from the perils of common sense. Only recently did they finally bring themselves to believe the seemingly absurd (but true) notion that the continents drift around, careening into each other like so many bumper cars, setting off earthquakes in the process.

And what about that rock from Mars carrying markings that looked like ancient fossils? Planetary scientists tell us it got chipped off the red planet by a visiting asteroid, then wandered about the solar system for 16 million years before plopping down in an Antarctic ice field. Just a few years ago, researchers didn't believe that chunks of Mars could get to Earth at all. Today, they estimate that 100 pounds of Mars rains on our planet each year.

The mathematicians have been the worst of all, turning common sense inside out with astonishing regularity.

First, they invented obviously nonsensical negative numbers. (What does it mean to have minus two apples?) Then, they discovered irrational numbers, like pi, that run on forever. According to legend, the nonsensical idea of irrational numbers was initially treated much like the nonsensical idea of a spherical Earth that revolved around the sun; scientists were ridiculed, or worse, for promoting such notions.

Today, mathematicians accept everything from imaginary and transcendental numbers to infinities that are bigger than other infinities, and twenty-dimensional geometries. And calculus—once denounced as an absurdity because it deals with ghostlike "infinitesimal" quantities—is now routinely studied in high school.

The unsettling truth is that nature doesn't give a hoot what humans think is "common sense," and great scientists have learned to accept this better than the rest of us.

Isaac Newton said flat out that his own theory of gravity was such a great absurdity that no reasonable person could believe in it because it relied on the notion of invisible influences spreading through empty space. Yet, it was Newton's laws that enabled NASA to send a spacecraft to Mars (and, for that matter, sent Mars rocks to Antarctica). The moral is: If it works, it's probably what Nature intended—sensible, or common, or not.

The late physicist Frank Oppenheimer used to get exasperated when people urged him to use his "common sense" and accept the world the way it was—whether the subject was science or social policy. Time and again, he'd remind them: "It's *not* the real world. It's a world we made up."

Seeing Stars

Sometimes, astronomy is unreal. Really.

Those spectacular pictures from the Hubble Space Telescope of exploding stars and nebulous gas clouds and stars emerging from dusty cocoons are all—to one extent or another—computer-enhanced. The rose-red Martian landscapes and candy-colored rings around Saturn don't reflect what you'd see in a telescope from your backyard. The images are processed, spiffed up, airbrushed, and painted like so many Hollywood stars.

The result is that people looking at the same sights through ordinary backyard telescopes often feel betrayed. They don't see enormous green and purple glowing clouds of gas. Saturn looks mostly black and white. This moment of truth is like Dorothy's discovery that the great and powerful Wizard of Oz is really just a guy behind a curtain gussied up with some crude special effects.

But is it deception, really?

When Galileo first looked through his simple telescope and saw mountains on the moon and moons around Jupiter, people thought he was seeing an optical illusion. His telescope, they said, was creating distortions. Anything you couldn't see with your naked eye, in other words, wasn't really real.

Of course, Galileo's detractors had hidden agendas. In a perfect world that was the center of a perfect universe, no one wanted to see imperfections on supposedly perfect heavenly bodies or, heaven forbid, satellites orbiting worlds other than Earth.

Today, we don't worry about the reality of the things we see through lenses. In fact, many of us walk around with lenses floating on our eyeballs or perched on our noses, the better to see the world.

Moreover, we know that the images we see in our mind's eye are exhaustively processed by our brains. The brain rights the upside-down images on our retinas, fills in blind spots, erases extraneous "noise" such as blood vessels and floating bits of fluff from our field of vision. The brain adjusts for motion, "corrects" colors, and puts things in their proper perspective.

In their own way, that's just what the Hubble scientists do. According to *Astronomy* magazine associate editor Robert Naeye, who put together a retrospective of Hubble treasures in his publication, the astronomers have to clean up "noise" in the images, such as cosmic ray tracks, and clear up distortions. "The astronomers have to do some processing to make them look spiffy," he said. "You have to do a little massaging to make them sing."

What's more, the Hubble images come down in black and white. They're snapped separately through various filters sensitive to different colors. The astronomers put them back together. That's also more or less the way the human eye works— three different kinds of color-sensitive cells "snap" images, which the brain reassembles.

So it would be difficult to argue that the Hubble images are somehow dishonest. After all, there's no such thing as an unprocessed, unfiltered image. For that matter, there's no such thing as an unprocessed sense perception, be it sound, touch, taste, or smell.

Color and smell arrive at the brain as electrical signals, which the brain "massages" into sensations such as "blue" or "freshly cut grass." As Galileo recognized, qualities such as color and smell "can no more be ascribed to the external objects than can the tickling or pain caused sometimes by touching such objects."

Whether or not we distort these perceptions when we "enhance" them with hearing aids or other similar devices is a moot point. Everything we sense from the outside world is created in our minds from heavily processed and filtered information.

And some objects in space really do appear downright tacky—even in natural light. While we think of the hot sun as the standard of "natural" light, cooler glowing gas clouds emit light in the same way as neon signs.

"They're nature's equivalent of going down a Las Vegas strip," said Hubble Space Telescope spokesman Ray Villard. "It gives us some very gaudy images." While the Hubble tries to keep its processed images as true to life as possible, the scientists haven't decided exactly what to call this "massaged" version of reality. "We have debated it here," said Villard. "If we try to say [the images are] true color, or natural color, that can be misleading. But if you say false color, people think you're trying to deceive them."

The truth of the matter is hard to explain: Yes, the Hubble images are processed, filtered, reassembled, artificially colored, and generally jazzed up. But they are no more jazzed up than the images we see with our own two eyes.

Calibration

When my eighty-year-old father started working out with a personal trainer, I knew I had to recalibrate the way I measured "old." Curiously, the process was quite similar to the way astronomers recently had to recalibrate the measuring sticks they use to find out the size, and therefore age, of the universe.

Calibration is central to science because it assures that instruments are measuring what they are supposed to be measuring. Call it a reality check. If something you're measuring doubles, you want to make sure that the measurement you take doubles, too. If a scale ranges from 0 to 100, you want to make sure that reality falls within that range. If your bathroom scale counts only one pound for every two you gain, it might make you happy, but you still won't be able to buckle your belt.

And if you think eighty is at the outer edge of "old," then there's no place in your scheme of things for an eighty-year-old lifting barbells; if you find such a person, you know there's something wrong with your measuring scheme.

Astronomers found themselves in this position when measurements suggested that the universe was younger than its oldest stars. Part of what came to the rescue was a recalibration of their measuring sticks. The rulers they were using to chart the distance to stars were apparently giving the wrong answers, thereby making the universe the wrong age.

The bigger the universe is, the older it is—because bigger means it's been expanding from its origin at the Big Bang for a longer period of time. If stars are farther away than astronomers thought they were, that means the universe is older than they thought it was (assuming, of course, that the rate of expansion is well understood—another controversial calibration point).

It would be easy to measure distances in the universe if stars came marked like lightbulbs—say, "60-zillion-watt star." If you knew the star's natural brightness, you could calculate how far away it was by measuring how bright it looked. The brighter the star looked, the closer it would be. To make this system work, however, you would need to know how bright the star was to begin with, because a very dim star close by could look the same as a very bright star way off in the distance.

To calibrate their cosmological rulers, astronomers try to find "standard candles," or stars that are marked like lightbulbs with clues to their intrinsic brightness. Using data from a now-defunct satellite called *Hipparcos,* astronomers at Caltech found that certain stars previously used as "standard candles" were 15 percent farther from the Earth than previously thought. This meant that their natural shine had to be about 20 percent stronger to appear so bright from so much farther away.

And that flaw in the measurement potentially throws off most other, more grandiose, measurements of the size and age of the universe.

Of course, one needn't be a cosmologist to run into the hazards of miscalibration.

Savvy chefs recalibrate cooking times depending on the altitude of the stove. Teachers use curves to recalibrate grades. A few years ago, the Scholastic Aptitude Test was recalibrated because falling test scores meant the average was no longer, well, average. A physicist friend even used his knowledge of pitfalls of calibration to talk his way out of a speeding ticket. He got stopped by a cop who clocked him at 48 mph. The physicist argued that his speedometer read 35; perhaps it wasn't calibrated properly. His ruse worked.

Physicists know about such things because they, like the rest of us, almost never measure anything directly. Since our brains don't come equipped with thermometers, we measure temperature by watching the height of a mercury column in a glass tube. We measure speed by the movement of a needle on a gauge.

The more esoteric the measurement, the farther removed it is from the thing you actually measure. If one calibration is wrong, the whole calculation can go out the window.

And just as I had to recalibrate what I meant by "too old" for barbells, biologists recently had to recalibrate what they thought was "too hot" for life. The bacteria they found living on the noxious fumes from deep undersea vents were thriving at temperatures thought far too searing for biological molecules to hold together.

No wonder some researchers are ready to believe that life existed on ancient Mars, or perhaps continues to exist on Jupiter's icy moon Europa. A little recalibration goes a long way when it comes to setting limits on life.

Word Problems

It's not hard to say what I love about physics. Or hard to spell, either. You won't find a term like polytetrafluoroethylene in a physics book. You'll never have to stumble over acrasiomycota or mycorrhiza or dinoflagellates. The words of physics tend to be short, simple, to the point: force, energy, momentum, mass, spin, black hole, dark matter. The trouble is, the words don't always mean what they say.

Take electric charge, for example. It comes in two varieties, positive and negative. But while an atom with a few extra electrons buzzing around its periphery is negatively charged, an atom with a shortage of electrons is positive. Go figure.

Or what about the term "accelerate"? Most people use the word to mean speed up. Physicists also use it to mean slow down, or even change direction. A biologist or a chemist probably would have made up a new term to describe these other kinds of acceleration, but physics likes to keep things simple. Alas, that can make it hard to tell whether you're coming or going.

Other terms are simply approximations that everyone knows you can't take literally. The term "spin," for example. Just about everything in the universe has it, including elementary particles. But some particles—like electrons—have no dimension. They are so small they are no more than points, giving them no proper "center" to spin around. Spin was as close as the physicists could come to describing a newly discovered property of particles that has no real description in everyday speech.

Some terms in physics even mean the opposite of their everyday definitions. Think of "perfect symmetry" and most people conjure up a snowflake or patterned tiles or the twin wings of a butterfly. To a physicist or a mathematician, those are examples of limited or even "broken symmetry." To them, an example of a truly symmetrical object would be a sphere, because no matter how you turn or move a sphere, it still looks the same as it did before you changed it.

You'd think that by now physicists would have learned to be more careful with the way they throw words around. But no, even new concepts get obscured with the same kinds of verbal ambiguity. In the 1970s, astronomer Vera Rubin confirmed that stars race around in galaxies so fast that they should fly right off into space. There simply isn't enough visible matter to provide the gravitational glue needed to stick the stars in place. There must be something else, unseen, and lots of it.

The still-unidentified source of this mass is generically known as dark matter. The problem is, the dark matter really isn't dark. If it were, it would cast a shadow—like the moon casts a shadow on Earth during an eclipse. Rather, it's transparent, like glass. You can't see dark matter not because it blocks light, but because light goes right through it.

Of course, sometimes physicists give names to new phenomena just for fun. The names for quarks are quirky: "strange," "charmed," "top," "bottom." One new hypothetical particle (a popular candidate for dark matter called the "axion") was named after a laundry detergent.

And why not? After all, putting the precise mathematical meanings of physics into images understandable to the human brain always loses something in translation. Pinning them

down with words is like trying to put your finger on a snowflake. The more precise you try to be, the more the meaning
melts away.

Perhaps that's why physicists don't feel bad about changing
the meanings from time to time. They were only approximations to begin with.

"One of the slippery things we do in science is we keep redefining words...as we learn," physicist Helen Quinn of the
Stanford Linear Accelerator Center told me.

One of the most egregious examples she cited was the word
"matter" itself. Textbooks define it as that which has mass and
takes up space.

That's good enough for chairs and planets and even atoms.
But what about particles like photons, which have no mass, or
electrons, which take up no space?

"We have changed the rules of the game," said Quinn.
"Now matter is defined as baryons and leptons."

"Lepton" and "baryon" might not be words everyone can
warm up to. But I'll take them over complementary schismogenesis any day.

Naming Names

Anyone who's been stuck with a stupid nickname knows how it can grate for life.

My mother thought it would be cute to call a baby by initials, and so I've been laboring under K. C. ever since—despite sporadic efforts to restore my good name(s).

My only revenge is that other members of my family are known variously as Wizzie, Boscar, Bidley, Mirp, and Uncle Do Do.

Alas, the same thing happens in science. Sometimes, with much more far-reaching results.

Take Big Bang, for example. The term originated with Fred Hoyle, who attached it to a bizarre theory of the origin of the universe—a theory so clearly wrong, he thought, it should be laughable. "Big Bang" was meant to be sarcastic; a pointed, if nasty, joke.

Of course, these days the Big Bang theory of the origin of the universe is considered the crowning achievement of cosmology. The theory has been so successful, astronomers say, it explains the creation of elements, the expansion of space, the structure of the universe itself; they even study its lingering echoes to determine the antics of newly forming atoms as much as 13 billion years ago.

Poor Hoyle! The term he heard as a discouraging word now is used with a straight (even reverent) face to describe a pillar of modern science.

In fact, jokes get taken seriously all too often in science. When Murray Gell-Mann saw a pattern in hundreds of species of subatomic particles that seemed to naturally group them into families of eight, he dubbed it the Eightfold Way. He insists now he didn't have the Tao of physics in mind. "It was joke!" he protests. But that doesn't stop people from associating Gell-Mann's groupings of quarks with Eastern mysticism.

You'd think physicists would have learned to restrain such clever coining of terms. But no. Now we have "the God particle." You might rightly wonder just what's so holy about this particular chunk of vibrating empty space (which is what the more appropriately named Higgs boson most closely approximates).

True, the current search for this hard-to-pin-down particle is considered extremely important to physics. But why "the God particle"? Because, explains Nobel laureate Leon Lederman, his publisher wouldn't allow him to call it "the goddam particle." It, too, was a joke.

Whatever his excuse, the name has "legs," as they say in Hollywood. Like K. C., it seems to be sticking around. I asked a physicist at MIT recently what he thought was the most important thing for people to understand about this Higgs boson. "It has nothing to do with God," he answered.

Some of these attempts by physicists to crack jokes have big unintended consequences. Like it or not, they can steer our imaginations in the wrong direction. Big Bang is a good example. The term implies some kind of explosion. So people often understandably ask: Where did the Big Bang happen? After all, explosions normally occur at a particular time and place.

But since the Big Bang created space and time along with the rest of the universe, the question has no meaning—except, perhaps, that the Big Bang happened everywhere, for all time.

Some silly names start off not so much as jokes as simple shorthand: "string theory," for example. According to string theory, all the particles in nature are really tiny loops of vibrating string. "Why string?" asked a friend. "Why not bagels?"

Because, of course, they're not really "string." They're mathematical descriptions.

What's more, these two-dimensional loops have expanded to include membranes and other complex objects. Still, "string theory" persists—suggesting that the universe is a big tangled ball of twine.

Einstein had an even worse problem with his theory of relativity—which he liked to call his theory of invariants. Relativity is fundamentally about the fact that the underlying laws of nature don't change no matter what; they are invariant to transformation. This means that appearances are often relative. But relativity is the theory's secondary, not central, teaching.

So to interpret relativity as "everything is relative" is almost precisely the opposite of its true meaning—which is that fundamental things are absolute.

Of course, the physicists—like parents—have to call their progeny something, I suppose. And people will call things (and people) names.

Still, I can say from personal experience that it's a lot easier to have a name that doesn't always need to be explained or interpreted.

(And by the way: It's Karen Christine.)

Context

In a high school science class recently, I gave the students some of my own articles from the *Los Angeles Times* to read as examples of science writing.

"Wow," said one of the students. "Your writing is really bad!"

It took me a while to realize it was a compliment. "Bad" was teenage for "good."

As always, context is everything.

In fact, one of the reasons that the discovery of zero produced such a profound revolution in mathematics is that zero allows you to put numbers in their place. The numeral 1 in the number 1,000,000 means something very different from the same numeral in the number 10 or 1,001.

In the same way, June marks the beginning of summer—if you happen to be in Los Angeles. In Buenos Aires, the same month marks the beginning of winter. The number 911 means "emergency" when dialed on a phone. But in another context, mathematician Ian Stewart points out, the same number could mean that you lost the lottery (the winning number was 872), or that you live on a long street. More recently, the number has become shorthand for the September 11 terrorist attacks.

Just as you may not recognize your dentist in the supermarket checkout line, scientists often have to strain to recognize scientific data out of context. The researchers looking at images of Mars through their camera on Pathfinder got a big surprise when the camera uncoiled to its full height, and suddenly the boulders looked smaller. The observers didn't have any context to judge the scale of the first images correctly.

With only one-third the gravity of Earth, geological forces on Mars also work in a very different context. Weaker gravity means that mountains on Mars can be much higher than mountains on Earth because gravity doesn't so readily pull them down. Canyons can be deeper because the walls don't so readily cave in.

The relief on Mars gets as high as twenty miles. "There's nothing like that on Earth," said Pathfinder project scientist Matthew Golombek.

Mars's more sedate internal dynamics change the geological context as well. Earth's continents slide about on enormous plates that float around on "a mushy mantle," Golombek said. Mars's terra is firmer. So tall mountains have more support.

Water flows differently in one-third gravity—which affects the way it eats away at rocks, tumbles and scours the landscape. When geologists try to reconstruct the biblical-scale floods that scarred the Martian landscape billions of years ago, they'll have to take this different context into account.

Indeed, everything flows differently depending on gravitational context. In zero gravity—for example, on the space shuttle— water doesn't flow at all but floats in spherical blobs. Even flames—which are burning gases—tend to form into spheres.

Some things even change their identity depending on their context. A cake crumb on the floor is dirt; the same crumb on the cake is food.

Other things change shape: The Earth is obviously round when seen from a distance—say, from the moon, or even an airplane. But a small patch of earth is flat.

Still other objects become invisible. Where do the stars spend their days? Up in the sky, but you can't see them in the context of the overpowering glare of the sun.

Gravity completely disappears in the context of falling. Indeed, some physicists dismiss gravity as a "pseudoforce" because it doesn't exist for something that's falling. That's why astronauts "float" in orbit around the Earth. Gravity hasn't gone away. Instead, the spaceship circles in a continual state of free fall, pulled toward Earth enough to keep it in orbit, but not enough to send it crashing to the ground.

Scientists often create confusion of their own when they use their in-house jargon out of context. A researcher may describe a result as "robust," but that doesn't mean it's big or heavy or strong; instead, it means the measurement has stood up to many tests. A result described as "nontrivial" is important or even profound.

Or perhaps "bad" would be a better word in this context.

Purpose

Bt what conceivable good is it?

That's a question frequently tossed at physicists who work on those murky frontiers of science where progress isn't always clear and practical purpose is the farthest thing from anyone's mind.

These scientists spend their time on exotic pursuits like cooking up new states of matter, mapping the shape of the universe, or fiddling around with mathematical strings in ten-dimensional space.

One might well ask: What for?

The traditional response of physicists has been to drag out Michael Faraday's legendary answer to the person who asked him what possible practical use could there be in his early explorations of electricity. Faraday reportedly replied, "Of what use is a newborn baby?"

Electricity is old news, of course, so physicists sometimes now update their story by telling us that all of modern computing and electronics is based on the understanding of the atom known as quantum mechanics—a bizarre set of behaviors that seemed about as technically promising at the time as the discovery of a new black hole.

But even quantum mechanics goes back to the early twentieth century.

What, one might well ask the physicists, have you done for us lately?

One rather surprising answer surfaced during the meeting of the American Physical Society in Long Beach last year: Einstein's theory of general relativity, of all things, guides the Global Positioning System that allows boaters, hikers, and drivers to know exactly where they are all the time.

You can't get much more esoteric than Einstein's general relativity. In essence, the theory explained the familiar force of gravity as the curvature of space-time. Like an elephant sitting on a water bed, heavy objects bend space-time into "gravity wells" that pull other objects in. The Earth can't escape the gravitational clutches of the sun, because it's sitting in the sun's strong gravity well and can't climb out.

Time is not immune to the pull of gravity either. So clocks close to strong gravitational sources run more slowly than clocks safely at a distance. This means that GPS satellites orbiting Earth tick away time faster than identical clocks on Earth.

(Curiously, Einstein's "other," or "special," relativity tells us that time slows down at the speeds the GPS satellites travel; however, the two effects do not exactly cancel out.)

Without general relativity, in other words, "GPS would fail," said physicist James Hartle at the meeting.

Of course, that's only general relativity's most obvious use. The theory also tells us that warped space-time bends light, so enormous concentrations of mass should act like lenses, forming images and distortions much like traditional lenses of glass or plastic.

Einstein thought the effect might never be seen, but today naturally occurring "gravity lenses" are the telescopes of choice for certain types of astronomy. Images of distant galaxies are bent into telltale curved shapes, indicating that a lot of unseen matter lies between them and Earth. Such observations help measure the amount of matter in the universe.

Taken to extremes, general relativity produces bizarre effects of severely warped space-time such as black holes. Not so long ago black holes were nothing but a theoretical nightmare; today they are taken for granted and have become a subject of intense scientific scrutiny.

"Black holes are no longer a theorist's dream," declared Hartle. "They have been detected."

As proof, he offered the work of UCLA astronomer Andrea Ghez, who has charted the orbits of stars around the black hole in our own Milky Way so convincingly that the hole almost seems visible.

And what good is a black hole, you ask?

Looking to the distant future, there are wild ideas around for harnessing the energy of black holes—even using them for interstellar travel. These are highly speculative, to say the least.

More immediately, however, black holes promise to pay off big as pristine laboratories where fundamental physics can be explored under extreme conditions.

Of course, one can't travel into a black hole without getting crushed into oblivion. But that doesn't rule out riding into those dark recesses with the help of mathematics and imagination.

Using these tools, physicists have already established that the known laws of physics break down in the hearts of black holes. Which means that black holes are almost surely portals to new physics, new understanding, new insights.

Who knows? They might even lead to a new generation of computers.

Or something far more interesting still.

Soul Food

Scientific research, above all, is supposed to be practical.

That's one reason certain members of Congress were refusing to fund a modest NASA satellite called *Triana* that would fix a steady gaze on Earth from space—giving the folks back home an opportunity for real-time, round-the-clock navel gazing.

Unlike other satellites that take patchwork snapshots, *Triana* would beam back a live image of the whole planet twenty-four hours a day. All the planet, all the time, available to everyone through the Internet and cable TV.

"Boondoggle," House Majority Leader Dick Armey (Republican–Texas) called it.

"Tripe science," said Representative Dave Weldon (Republican–Florida).

It didn't help with Republicans that the idea was hatched in the head of then Vice President Al Gore, during a dream, no less. The image of our small blue marble floating alone in space, Gore said, would inspire people to take better care of the planet.

But is the purpose of science to be inspirational? Since when is science supposed to be soul food? Isn't its aim to produce the kind of technology that puts real food on the table?

After all, taxpayers have a right to want their money's worth. So whenever a new discovery is announced, politicians (and newspaper editors) are forever asking: What good is it? Of what practical use?

Scientists, in turn, have become adept at ticking off practical benefits of even the most esoteric research. During the early days of the space program, we heard a lot about spinoffs like Tang and Mylar. These days, we're more likely to hear about space-age drugs and computers, or medical imaging technology like PET scans—brought to us courtesy of a particle of antimatter called the positron.

Rarely, however, do people talk about the emotional payoffs of science. Like art museums and national parks, science fills deep philosophical yearnings. One doesn't have to look far to find these emotional fruits of science. Consider how much our feelings have changed about our place in the universe since Galileo discovered the moons of Jupiter, showing that not everything orbits around Earth; since Newton found that the laws of gravity that make apples fall on Earth are the same everywhere in the cosmos.

No longer could people view the dance of the stars and planets as a performance put on specifically for us. We are part of the action—and a small part, at that.

These days, astronomers tell us that perhaps 90 percent of the universe is made out of matter completely unlike the stuff of which we ourselves are made. No one knows the exact nature of this "dark matter," but whatever it is, it's alien to us. Or more accurately: We are the aliens in a universe constructed mostly of other kinds of material.

Scientists as well as politicians sometimes underestimate the emotional pull of science. Frank Oppenheimer—banned from physics by the politics of the McCarthy era in the 1950s— taught high school for some time in the small town of Pagosa Springs, Colorado. Teenagers, he thought, would be naturally interested in cars, so he'd take his classes to the junkyard to study the physics of auto parts.

He was surprised, one day, when the students complained. It was all well and good to learn about mechanics, they said. "But we want to know about the stars!"

No doubt, the same yearnings prompted millions of people to stay tuned to their TV sets for days following the adventures of a small rover called *Sojourner* on the surface of Mars a few years ago. The Pathfinder mission was, above all, a sentimental journey.

Triana eventually grew from a mere dream to a full-fledged scientific mission with a practical job in space. In addition to providing a permanent "mirror" on ourselves, it would keep an eye on environmental shifts and climate changes.

Still, one shouldn't underestimate the power of dreams. Most scientists, at some level, are dreamers. Practicality can get you to the next step in science. But only dreams can guide the next great leap.

Indeed, former NASA chief Dan Goldin notoriously scolded particle physicists for forgetting how to dream—for being, in effect, too practical. "If we don't dare to dream, we won't find anything," he said in a speech at Fermi National Laboratory outside Chicago. Dreams, he said, are "how the most exciting science happens."

This sentiment was perhaps best expressed by Fermilab's own founder, physicist and sculptor Robert Wilson. When Wilson was trying to get money out of Congress to build Fermilab in the 1970s, he was asked to explain the practical purpose of the new accelerator. How would the knowledge that came out of smashing atoms help secure the national defense?

Wilson responded that the new accelerator would have no value in that respect. Instead, he argued: "It only has to do with the respect with which we regard one another, the dignity of people, our love of culture. In that sense, this new knowledge has nothing to do directly with defending our country—except to help make it worth defending."

A lot of scientists would second that emotion.

Lies

"You always have to lie a little to tell the truth." A first-rate popularizer of science, the late MIT physicist Victor Weisskopf, once told me that was the secret to explaining science in nontechnical terms. He should know. His enchanting book *Knowledge and Wonder* has been seducing laypeople into the world of science for decades.

And so I confess: Week after week, I tell what I hope are well-considered lies. I write about electric currents that "flow" through wires like rivers—even though I know they do no such thing. I talk about elementary "particles" and "forces" even though physicists have only the foggiest notion what those terms mean. I toss off words such as "time" and "gravity," knowing that, to some extent, their very existence is illusory.

The lies are simplifications that ease the way to the central point. The art is knowing the difference between acceptable— even necessary—lies and oversimplifications that slide into mistakes. Sometimes, it's hard to tell which is which, and even scientists frequently don't agree.

I vividly remember the first deliberate lie I told in a science essay. I was writing in Time Inc.'s *Discover* magazine about the ways seemingly small differences can sometimes produce enormous effects.

The elements neon and sodium, for example, have starkly contrasting characters. Neon is what chemists call a noble gas.

It is so self-contained and standoffish that it won't react with anything. Sodium, on the other hand, is a highly reactive metal.

And yet, the only difference between the two, I wrote, is one lousy extra electron in sodium's outer shell. Neon has ten electrons to sodium's eleven.

To be honest, I stole the example from Weisskopf (who also told me that the only sin is if you hear a good idea and you *don't* steal it). And strictly speaking, it is wrong. There's something else different about neon and sodium: Sodium also has an extra proton in its nucleus. However, atomic nuclei don't affect an element's chemical character. And as the mathematicians like to say, a difference is a difference only if it makes a difference.

Was pushing that proton under the rug an acceptable lie? Or a misleading representation?

More recently, I wrote a column about water that contained both an acceptable lie and a misleading misrepresentation—and received a dozen letters from chemistry teachers in response.

In describing the marvelous and mysterious chemical properties of water, I said it was the only substance that can exist in solid, liquid, and gaseous states at the same temperature. That sentence should have read: Water is the only substance that is solid, liquid, and gas at temperatures and pressures normal on the surface of the Earth.

Actually, all substances have a so-called triple point, some combination of pressure and temperature at which they can exist in all three states at once. But it would take extreme temperatures and/or pressures to reach most triple points.

Because the essay's focus was water's relevance to life, I omitted the second half of the thought. In retrospect, that was a misleading misrepresentation.

The lie was my statement that water is the only substance that expands when it freezes into a solid. Actually, the metal alloy used to set type also expands. That embellishment seemed to distract from the central point, however, so I chose to leave it (like sodium's proton) out. In a case like this, less seemed more. There's a limit to the number of qualifications you can pile on a point without muddying its meaning.

Honest descriptions of scientific ideas will always remain elusive, mainly because an accurate depiction lies somewhere between the crisp clarity of equations and the fuzzy familiarity of metaphor. Even equations, when you get right down to it, don't describe the "whole" truth very well, because truth has many facets. Only a limited slice of reality can be accurately described by any given set of natural laws.

Or as the founder of quantum mechanics, Niels Bohr, reportedly put it: "There is an uncertainty relationship between truth and clarity." In physics jargon, that means the closer you get to pinning down truth, the more elusive clarity becomes, and vice versa.

At times, being clear and telling the whole truth may be mutually exclusive.

PART II

Stuff

Surprises

You just never know what the universe is going to be up to next.

Why, just this month, a big glob of matter sitting out in space bent light like a giant lens, bringing into focus a far-off baby galaxy just coming into being.

Not a month before that, the black hole at the center of our galaxy let out a loud belch of X rays, the best evidence yet that such a monster was feeding there voraciously—proof from the belly of the beast, as it were.

This is only the latest in a long string of surprises; it sure makes you wonder what else the universe has in store.

A paltry few hundred years ago, people believed there was nothing beyond what they could see with their eyes. How unnerving when the microscope introduced us to the teeming population of microbes that lives within (and on) our skin! How pivotal in the history of humankind!

"Who would have dreamed," writes Arthur C. Clarke, "that a tube connecting two lenses of glass would pierce the swarming depths of our being, force upon us incredible feats of sanitary engineering, master the plague, and create that giant upsurge out of unloosened human nature that we call the population explosion?"

Up until the start of this century, people thought that molecules were useful only as models because, of course, everyone knew that molecules could never be seen. Then Einstein showed that they not only could be seen, but had been seen, knocking about plant spores floating on water.

In short order, people learned not only to see molecules and atoms, but also to look inside. Ernest Rutherford discovered the atomic nucleus when he bombarded gold atoms with particles streaming from radioactive rocks. Most of the particles passed right through, but some—surprisingly—were scattered backward. Rutherford wrote: "It was quite the most incredible event that ever happened to me in my life. It was almost as incredible as if you fired a 15-inch shell at a piece of tissue paper and it came back and hit you."

People had also assumed, with good reason, that it was impossible to know the composition of stars, since it was hardly possible to go and collect a sample. Then shadows were found in starlight that spell out a complete list of ingredients—a quantum mechanical bar code that reveals everything but their price.

Long before Oprah, stars were spilling the intimate details of their lives to all who cared to listen.

In fact, it was by decoding starlight that astronomers discovered (surprise of surprises) that 90 percent or more of the matter in the universe is unseen and perhaps unseeable.

Many discoveries have been so surprising that people didn't believe them, even when the evidence (like the jostling of plant spores) was right before their eyes.

When Marie Curie first explored radioactivity, she speculated that the phenomenon might be evidence that atoms were actually disintegrating. This was inconceivable at a time when atoms were considered the indestructible building blocks of nature. Atoms—even more than diamonds—were forever. Her idea seemed so bizarre that she decided not to publish.

True, some discoveries are not so surprising. Recently, studies conducted with positron-emission tomography (PET scans) revealed that the brains of teenagers are completely unlike those of adults. "Duh!" a parent (or teenager) might be tempted to say.

But the very fact that positrons, which are a form of antimatter, could be used for medical purposes is pretty darned surprising—not to mention the fact that there is such a thing as antimatter at all.

We don't even know what it's possible to know. Today, some scientists say we can't know what lies beyond (or before) the Big Bang; others think they know how to look for evidence.

String theory has been dismissed as so much pretty mathematics since it can't be experimentally tested.

Yet just last year, the University of Chicago hosted a symposium on, yes, experimental tests of string theory.

In *The Unexpected Universe*, naturalist Loren Eiseley tells of coming upon a spider in a forest, spinning the sticky spokes of the web that extend her senses out into the world. Just so, humans with their scientific senses have spun a web that reaches far beyond our ears and eyes. And like the spider, we lie "at the heart of it, listening."

Yet Eiseley is even more impressed at what the spider cannot perceive. "Spider thoughts in a spider universe—sensitive to raindrop and moth flutter, nothing beyond....What is it we are a part of that we do not see...?"

Whatever it is, it's sure to be unexpected.

Roots

How far back can you trace your roots?

A chance meeting between Mom and Dad? The *Mayflower*? An early Egyptian king?

Some people trace their family trees obsessively, rooting around for the sources of a peculiar down-turned nose or up-turned eyebrow; for perfect pitch or debilitating neuroses; for inspiration, or inheritance, or both.

For better or for worse, our roots anchor us in place. Everything else is the tip of the iceberg, proverbial and otherwise.

Icebergs, as the *Titanic* found out the hard way, have deep and often invisible roots. The roots support the tip. If it weren't for the root, the whole thing would sink.

This was the discovery that reportedly led Archimedes to run naked down the streets yelling "Eureka," still dripping from his bath: It's the roots of things that enable them to float. It's what's beneath the surface that holds everything up—whether it's a rubber duck floating in a tub or an ice cube floating in a glass.

So it makes sense that we spend a lot of time digging around for roots. Roots support, nourish, anchor, explain.

Consider a mountain. How tall can it get? It all depends on the root.

Mountains, like icebergs, float on the denser material (the Earth's mantle) underneath. The taller the mountain, the deeper the root, and even though you can't see the root of a mountain, you can detect it indirectly. The massive root of the Himalayas, for example, has been measured by its gravitational pull on a sensitive pendulum bob.

You might conclude from this that most of what we know is but the tip of some kind of iceberg—just a small peek at what lies beneath. Like the symptoms of disease. Like the whitecaps on water. (Did you know that underneath the crest of every ocean wave is a deep circular current?)

It seems to be so.

Indeed, the very matter that makes up Earth and stars and sky appears to be nothing but froth on a deep ocean of dark matter whose swirling currents carry along everything else. We don't yet know what dark matter might be, but it's responsible for the shape of the galaxy, the fate of the universe at large. All the twinkling stars and glowing galaxies, cool moons and pastel planets (as well as whatever life hangs out on them) are but the grin on this sly Cheshire cat; the tiny beacon of the lighthouse hiding huge outcrops of rock; the silly bit of wagging tail on the big, dark, mute dog.

Even atoms have structural roots. Take carbon, my personal favorite. Life is based on carbon because carbon has four outer electrons that stick like little strips of Velcro to atoms next door, allowing the atoms to form long and elaborate chains. Think of these attachment points as outstretched hands.

What gives carbon this marvelous property?

The electrons are anchored by the positive attraction of protons deep in the nucleus. Unlike the root of a mountain or iceberg, the root of an atom takes up but a tiny amount of space. But in mass, it accounts for nearly all of the atom's heft. The root is also the core.

That's why roots have such great explanatory power. They tell us why things are the way they are. The underpinnings not only of atoms, but also of crime, poverty, war. We call them root causes for a reason.

Similarly, cosmologists study the origins of the universe not merely because they are curious about how it came to be (and they are), but also because origins are instructive. Why is our universe constructed of gravity and electricity, electrons and quarks, three dimensions of space and one of time?

Is it some quirk of cosmic fate? A chance event, like your mother meeting your father? Or is it the inevitable consequence of the inherent nature of physical law? More like an arranged marriage?

Does our universe itself have roots in other universes?

A deep enough look at the roots of time might help us find out.

And if we could understand the roots of life, the origins of the first living beings, the forces behind evolution, we could better understand ourselves. We might even be able to predict what kinds of life could exist on other worlds, and how to contact it.

On this planet, at least, we know that life's roots go deep, roughly four billion years.

It's hard to find anything that doesn't depend for its sustenance on hidden roots: volcanoes, rivers, hair, teeth.

Kill the root and destroy what flowers on the surface. That would include, among other things, culture. For ideas also spring from roots.

So mark well what lies beneath: the water, the words, the laws, the ground.

Don't neglect your roots. You can't leave home without them.

Stuff

M y editor hates it when I use the word "stuff." Since I write about physical science, he thinks I should use more scientific (or at least precise) terms. Like "atoms" or "energy" or "matter."

But often, "stuff" is the only word that describes what I mean.

Take the stuff we call empty space, for example. The vacuum is a major player today in physics. It has shape. It has properties. It can freeze or melt. It has a history, and physicists think it has changed significantly over time. And yet, by definition, it's nothing. Not energy, not matter. Just stuff.

And what about mind? Is it matter? Clearly, the brain is composed of matter in the form of neurons and the atoms and molecules that make them up. But just as clearly, there is something more involved.

After all, every atom in the average body is replaced every half a dozen years or so. What, then, is a person? What and where are thoughts and memories to be found?

A concept like mind—as opposed to the physical brain—cannot be described as energy or matter alone. Instead, it's an exotic kind of stuff, something like a flickering candle flame. Even though the exact molecules of wax or photons of light in the flame are constantly changing, the form stays remarkably consistent. It's an evolving pattern of matter/energy stuff—

real enough to touch, even though you can't quite put your finger on it.

Attempts to assign labels more specific than "stuff" are often misleading because so many phenomena combine aspects of both. What do you call a fire, for example? Or an egg in the process of cooking? Some curious alchemy of matter and energy takes place that transforms liquid yolk to solid egg salad, solid wood to light and carbon dioxide. What do you call the cauldron of electrified particles and nuclear energy that fuels the shining of stars?

Carl Sagan once said we were star stuff—not star matter, or star energy. It takes more than matter or energy alone to make exploding stars or chemical elements. It takes the energy locked in matter, and the matter forged from energy, to arrange and secure the particles in place.

It was Albert Einstein, of course, who proved that matter is only a frozen form of energy, and matter and energy are different forms of the same basic stuff. Fires and nuclear explosions and the slow metabolic burning of calories in our bodies turn matter into energy; plants take energy from the sun and turn it into matter. You can add some energy to common soot and cook up the toughest stuff in the universe—diamond.

Einstein also showed that space and time coexisted in an inseparable stufflike state called space-time, and it does not take physics to see why. When it was midnight on December 31, 2000, in L.A., Tokyo was already well into the new millennium. You can't tell time without knowing where you are. And you can't know where you are unless you also know the time—because every second, everything in the universe (including us) is sweeping through space at breakneck speeds. Every millisecond, your position in space changes by hundreds of miles.

There is no where without when, no when without where.

"Stuff" is the only word to use when trying to describe unlikely matings like mind/brain, space/time, matter/energy. In fact, it's a good word to use for just about anything that has definite properties but can't be easily defined. Magnetic fields and forces. Or mathematical objects, like prime numbers, or pi.

Emotions, ideas, and poetry are stuff. Maybe that's why a chemist friend who also writes poetry freely sprinkles his writings with the word "stuff." He, too, has run into reluctant editors, who expect more from a Nobel laureate. He protests: "But it is stuff!"

The stuffiest of all sciences is a field called condensed matter physics—literally, the physics of stuff. It deals with liquids, glasses, metals, fluids, gases, ceramics—the crowd behavior of units of matter and energy congregating in space and time. Everything from sugar cubes to biological systems, miracle glue to human tissue, liquid helium to the structure of empty space. These days, it's where a lot of the most interesting research is taking place.

Even the stuff of dreams is coming from condensed matter: UC Berkeley's Marvin Cohen has dreamed up a new kind of stuff that's tougher than diamond. Various experimenters are now trying to condense his dream into concrete reality. It's very hard stuff.

Surfaces

People are always telling us that surfaces don't matter. It's what's on the inside that counts. Skimming the surface, they say, is superficial.

If you ask me, however, the inside is vastly overrated. Outsides are where the action is. Think walls, borders, ceilings, membranes, crusts, skins, doors. Surfaces are where the rubber meets the road—both literally and figuratively. Edges are a lot more than frills.

This is hardly news to the people of L.A., living, as they do, on the uneasy border of several rather large continental plates. The landmasses on either side of the San Andreas fault are traveling at cross-purposes—one creeping toward Canada while the other makes tracks for Mexico.

But it's what happens where the two slip past each other that makes life shaky for those who live upstairs, sending shock waves through the earth to rattle our glassware, our freeways, our nerves.

Earthquakes are only the most obvious example of action at the edges. The walls of cells, like the walls of castles, are where you'll find armies of molecular sentries standing guard, ready for action, deciding what comes in and goes out. The surfaces of water and earth are places where life likes to bloom.

Surfaces, after all, are where things make contact, including land, sea, and sky. So it comes as no surprise that nearly all the life that populates our planet makes its home on or near the thin margins of its crust.

Surfaces also tell stories.

The geological history of the planet is written in the wrinkles on its face—the great slashes where rivers run through it, the giant pockmarks where asteroids have come to call, the ashy mounds where hot rock has blasted right through, depositing mountains. Overactive hot spots, like the overactive glands of adolescents, break out on the surface of the skin, occasionally squirting out—among other things—diamonds cooked deep inside the earth.

Everyone knows that people wear their emotions on their skins: We turn red with embarrassment, white with fear. But stars, too, show their colors to broadcast their ages, their temperature, even their compositions. The spots on their surfaces tell of fierce magnetic storms roiling underneath.

There are all kinds of surfaces. Horizons, for example, are the surfaces of what we can see. The Earth's horizon told early sailors that our planet couldn't possibly be flat. Ships approaching it sank slowly into invisibility, trailing their masts behind them, rather than disappearing precipitously over some abrupt edge.

In the same way, it is the horizons of black holes that broadcast the existence of these great voids in space-time. Matter, as it falls in, should skid around the edges, stirring things up, sending out a last gasp of gamma rays before retiring into permanent oblivion. Does the universe have a horizon? Is there a surface that enfolds everything? An edge to space and time? The answer, curiously, is yes and no.

Space does not come to an abrupt end, although it may well curve around on itself like the surface of the Earth, finite but unbounded.

Time, however, is a different story. Since light travels at a finite speed, we cannot see light signals that would take longer to reach us than the age of the universe. So, for all intents and purposes, there is a wall in time—the edge of the observable universe.

Another interesting cosmic wall marks the moment—when the universe was about 300,000 years old—that neutral atoms were first formed by electrically negative electrons joined with electrically positive protons. Light was then finally free to separate from matter, making the universe transparent.

Remarkably, that light still lingers, imprinted with information about the universe at these earliest moments. Astronomers call it the "surface of last scattering" because it marks the "surface" in time when this light last interacted with matter. This surface, called the cosmic microwave background, promises to tell scientists much about the shape and composition of the universe at age 300,000.

Of course, it's true that some surfaces are simply nice to look at. Soap bubbles get their colors from thin films that reflect light off closely separated surfaces.

Sometimes, thank goodness, beauty really is skin deep.

Wind

Sometimes I sit and watch the wind have its way with the tree outside my house, blowing the leaves into billowy curtains, making them shimmer like sequins on a dance dress.

Of course, you can't see wind. You can only see what it does to things: twirling scraps of garbage into miniature tornadoes; slinking through flags or snapping them to attention; fashioning coiffures of clouds.

We think of wind as something that happens on our home planet: the trade winds that wrap the Earth in swift ribbons, the jet stream, the Santa Anas.

But the universe as a whole is a remarkably windy place.

Jupiter's fierce winds shear against each other as they shift along with latitude, setting up huge standing hurricanes, some the size of Earth. These storms are visible as red and white oval "spots." The hot winds of Venus blow at hundreds of miles per hour. Even gentle Mars has winds strong enough to stir up tiny dust devils on the planet's surface.

There is complicated "weather" in galaxies, too, writes British astronomer Sir Martin Rees, "churning up the interstellar gas and recycling it through successive generations of stars— this is how the atoms from the periodic table are built up from pristine hydrogen."

The winds of stars even liberate the atoms that compose ourselves, cooked, as they are, in the crucibles of stars.

Tightly packed clusters of stars can be particularly windy; so many stars in close quarters engage in a fierce gravitational tug-of-war, tearing gas off each other's atmospheres.

Some stars are so luminous that they literally blow their tops, the pressure of light lifting off layers of atmosphere so rapidly that the star loses the mass of Earth every few days. Live fast, die young.

When stars explode, escaping winds can travel at one-tenth the speed of light, twenty to thirty thousand miles per second. Indeed, the winds of dying stars create some of the most beautiful objects in the sky: planetary nebulae. When the fast winds from the stellar explosion plow into slow winds lingering from the star's previous exhalations, they drape the star's spent cinder in bulbous glowing shrouds.

Our own sun will one day take its place among this photogenic assemblage. For now, it has its own, albeit tamer, electrically charged wind, sometimes slung toward the Earth by powerful magnetic fields. Descending at Earth's north and south poles, the winds light up the sky in ghostly pink and green auroras.

Scientists can use winds to extract information from nature. After all, a wind is caused by a pressure gradient that makes particles of matter move from one place to another, which is why you can create "wind" by sucking through a straw or blowing through a flute.

Conversely, the detection of wind points to the existence of a difference. Thus, it was the lack of a detectable "ether wind" that helped bury the long-cherished idea that an invisible light-carrying substance pervaded all space. Since Earth moved through the ether, scientists reasoned, it must set up a "wind" as it plowed through. But no wind was ever found.

Today, physicists searching for dark matter are doing much the same experiment, only searching for the WIMP wind.

WIMPs (weakly interacting massive particles) are a hypothetical species of dark matter, attracted to our galaxy like moths to a flame but so ephemeral they are all but impossible to perceive other than through gravity. However, as Earth-bound detectors move through this cloud of WIMPs, they should feel an ever-so-slight (you might say wimpy) breeze.

Once you start thinking of wind as matter in motion, you can see it almost anywhere. A galaxy is a spiral "wind" of stars; smoke going up a chimney a sinuous wind of soot; matter falling into a black hole a wind that goes one way.

The universe began when the Big Bang blew space and time into being; that wind of ever-expanding space still blows, sweeping distant galaxies farther away from us, maybe even picking up speed.

There are winds so slow we barely perceive them: the aroma rising from roast beef (a "wind" of food), glaciers (a "wind" of ice).

Others are buried out of sight: the whoosh of hot mineral-rich smoke jetting out of deep undersea vents; the wind of blood that blows through your heart.

Living things make all kinds of wind—breathing, singing, crying, barking, tweeting. Bad breath is a foul wind. Whispers are winds from the heart.

Crowds funneling through a doorway make a "wind" of people. Traffic zooming along the freeway makes a "wind" of cars. All winds collide.

Winds also carry things with them: smells, secrets, radioactive fallout.

I read somewhere that after the World Trade Center attacks, debris traveled in the form of powder up the streets of Manhattan at fifty miles per hour—a wind of concrete and souls.

That windchill factor froze me to the bone.

Clouds

Got your head in the clouds? Well, why not?

It's hard to imagine a more interesting place to be. In fact, it's hard to imagine just where you could be in the universe and *not* be in the clouds.

Even familiar clouds are among the few phenomena that truly (almost literally) qualify as incredible. "Suppose," writes crystallographer Elizabeth Wood in *Science from Your Airplane Window,* "you had the task of trying to make [a] blind person believe them." Suppose, indeed!

Floating on thin air like foam on cappuccino, a good-sized cumulus can easily weigh 500,000 pounds. Yet for all their bulk, clouds are easily the most "uninhibited of all natural phenomena," wrote the late Guy Murchie in *Song of the Sky:* "Nothing in physical shape is too fantastic for them."

That's because clouds are wind made visible, skywriters that pen their messages in tiny drops of water (or sometimes ice), telling the world below (and also above) where currents are rising, racing, colliding.

Clouds, as Murchie so aptly put it, are "the architecture of moving air."

Not all of a cloud is necessarily visible. The curly crown of a cumulus is only the very top of a tall cloud tower that reaches clear down to the earth; air heated on the ground puffs up and rises, eventually cooling enough for the water held inside to condense. The cloud itself is but the ice cream perched on a long, invisible cone.

The creation of that high-flying ball of fluff is curiously similar to the process that transforms a kernel of corn sitting in hot oil into an edible puff. But then again, what is popcorn but bite-sized clouds?

For that matter, what is a dandelion but a cloud of fluff? My daughter's cat but a cloud of orange fur? Down pillows but clouds of feathers? An actress's puffy lips but a cloud of collagen? Smoke, insects, and perfume all make their way around the world in clouds.

Even an atom is but the tiniest nut of a nucleus entirely engulfed by a cloud of electrons. Until recently, physicists even hunted down elusive subatomic particles by watching for the wakes they left in "cloud" chambers.

There are clouds everywhere we look, and even some places we'd perhaps prefer not to. Computer pioneer Alan Kay, creator of many of the computer interfaces we now take for granted, described the problem of talking with your computer to me recently in terms of "thought clouds."

"When you say something to somebody you're actually pointing to something in your thought cloud that you hope is in the other person's thought cloud too," he explained. The problem is, people have trouble connecting with a computer's thought cloud (and, of course, a computer doesn't have a thought cloud until a person puts it there). The use of windows that appear and disappear, said Kay, made it possible "to put the computer's thought cloud on display." Using a mouse, he said, is "literally how you point to your computer's thought cloud."

If that isn't spacey enough for you, consider space itself, which is easily the cloudiest realm in the cosmos. Stars turn on in clouds, and also die in them—sometimes leaving behind garishly glowing nebulae (but another word for cloud). A star that dies in a massive explosion is a cosmic-scale cloudburst.

Other nebulae forever remain dark, littering up the galaxy. Astronomer Steve Maran, in *Astronomy for Dummies,* calls these dark nebulae "the dust bunnies of the Milky Way."

A galaxy is a cloud composed of stars instead of drops of water. In fact, the two sidekick galaxies of the Milky Way aren't even called galaxies, but rather the Large and Small Magellanic Clouds. Jupiter's famous bands are really stripes of pastel-colored clouds. A few years ago, a group of astronomers announced they had discovered a huge cloud of antimatter hovering over the center of our galaxy.

It's hard to imagine how clouds got such a bad name—dismissed by some as so much fluff, cursed by others as foreboding.

Clouds do have a dark side, of course, figuratively as well as literally.

Ironically, for example, clouds are the biggest factor obscuring the current understanding of global climate change. Some clouds act like beach umbrellas, keeping the planet shady and cool; others are like blankets that hold heat in. Which will dominate in a warming climate? No one knows.

Those thick, billowy cumulus—no doubt, the "marshmallow clouds" John Lennon had in mind—reflect most of the sun's light back into space. When you see them from a plane, Elizabeth Wood points out, the tops are white. Even from the ground, you can sometimes see the sunshine leaking over the edges, giving dark clouds a silvery fringe.

So some clouds really do have a silver lining. Only it's not hidden inside; it's sunny-side up, visible to all who get high enough to see.

Patterns

Physics is simple. In fact, it never ceases to astonish me how, with a few simple ideas from physics in your suitcase, you can travel from your own backyard to the most exotic realms of the cosmos, feeling equally at home in both inner and outer space.

Consider, for example, the familiar way water evaporates from a puddle, or freezes when the temperature drops. The universe began, cosmologists say, when the vacuum of empty space went through a very similar transition, from a melted state into a frozen one. Superconductors carry electricity friction-free because they are "superfrozen" to the point where they behave like a single large atom. At the other extreme, the glowing gases of auroras and stars are simply matter in what you might call an even more melted state, in which electrons are knocked loose from atoms.

This simple business of freezing and melting explains everything from the "quark soup" (melted protons and neutrons) that made up the early universe to liquid helium's ability to flow uphill (because the "liquid" is frozen into a single quantum state).

Or consider the harmonics of a bottle of beer. Blow over the top, and you can make a series of different sounds depending on how hard you blow and how much beer is left in the bottle. And lo and behold, it is by analyzing a very similar set of harmonics set up by the sloshing of gas and light in the early universe that astronomers have been able to put their ears to the cosmos, listening in on its babblings from the first moments of time.

Among the most beautiful of pervasive, simple phenomena is interference: what happens when waves alternately march in and out of step, adding up and canceling out.

In the thin skins of soap bubbles, these alternating bands show up as serpentine ribbons of color, each pastel shade appearing in the place where the thickness of the skin causes those particular wave lengths of light to march in step. Holograms are created by interference patterns encoded with the information needed to make three-dimensional images hover in thin air.

Interference can vastly magnify patterns, making the invisible visible. For example, light waves are far too small to see individually with your eyes, but the interference patterns shine through clearly in the iridescent colors of opals, oil slicks, and butterfly wings.

This magnification effect makes interference patterns uncannily useful. Your 767 en route to Maui relies on laser-produced interference patterns to inform the plane's inertial guidance system of even the slightest changes in position. Rosalind Franklin used interference patterns produced by X rays shining through crystallized DNA to first "see" the spiral structure of the molecule, and today most studies of materials rely on interferometry in one way or another.

So does most astronomy. Radio astronomers have long used interference patterns of radio waves to see into the distant sky—sometimes using arrays of telescopes spread across the globe, combining waves from each to create a single vastly improved image.

Today, virtually all major optical telescopes are adopting interferometry as well, including Mount Wilson, the premier Keck telescope on Mauna Kea in Hawaii, and the European Southern Observatory's Very Large Telescope in Chile.

One of the most exciting ways astronomers are using interference is to try to catch gravity waves from cataclysmic events in space—the collision of a couple of black holes, for example. Such a horrendous clash would shake up the four-dimensional fabric of space-time itself, sending out ripples that should reach our shores.

Alas, space-time doesn't get shaken up easily, and these ripples are almost inconceivably weak. Thus, the Laser Interferometer Gravitational-Wave Observatory consists of two enormous interferometers on opposite ends of the country, each strung with four-kilometer-long laser beams. Plans are in the works to put even bigger interferometers into space— probably the only way to see Earth-like planets.

Any rhythmic pattern can produce interference patterns: ripples caused by stones dropped into a pool of water; the threads in moiré silk; the overlapping struts of picket fences. Musicians listen for them (they sound like subtle throbs) to tune their instruments.

Indeed, because only wavelike phenomena can interfere, it was the discovery of interference patterns that proved that light was a wave—and later that elementary particles, too, have wavelike alter egos.

From the fringes at the edges of shadows to the "electron holograms" described in a recent issue of *Physics Today*, interference patterns are ubiquitous—synchrony and asynchrony playing off each other in a constantly unfolding fugue, poetry and utility in one simple, easy-to-carry package.

Put that in your suitcase, and I imagine it will take you on a trip somewhere interesting really soon.

Recycling

D on't pick that up!" our mothers used to scold. "You don't know where it's been!"

Our mothers wanted everything their children encountered in this world to be somehow shiny and new, untouched by other human hands, unsullied and pure.

These days, of course, almost everything we pick up has had some previous incarnation somewhere else: We walk around on rubber soles that were formerly parts of tires, write on stationery that was once paper money (or even old jeans), wear fuzzy fleece jackets recycled from plastic soda bottles.

Even yesterday's banana peels and coffee grounds get born again as next spring's begonias.

Like Buddhist souls, everything seems to have at least one past life.

One of the great surprises of twentieth-century science was the discovery that what is true of fleece jackets and rubber soles is true of nearly everything in the universe.

Consider the solid ground on which you stand. It's made of rock that's been recycled many times, melted down and remade in that great churning pressure cooker that simmers constantly beneath the deceptively cool surface of the earth. One day, your backyard will sink back under the ocean, to be cooked down as scrap, perhaps to emerge again as the outpouring of some volcano.

Or take a deep breath. Do you know where it's been? Almost certainly, it has been inhaled and exhaled by someone else. According to some estimates, a good number of the molecules we suck up in every breath almost certainly spent some time in the lungs of Cleopatra or Napoleon.

And so it goes. No matter where you look, the stuff of Earth has been there, done that, somewhere, somehow, before.

One of the most spectacular products of recycling is our solar system itself—that neat assemblage of fierce central sun with its well-behaved family of nine circling planets. Where did it come from? Where has it been? Where were the elements that make up our sun and planet before they were part of us?

Physicists have known for some time that nearly all elements heavier than helium are cooked inside stars—the heaviest in exploding stars, or supernovae. Therefore, everything we are today must at some point have been "star stuff," as Carl Sagan so famously put it. Our sun and planetary system condensed like globs of cold soup out of the hot, gaseous breath of some now-long-gone supernova.

The question is: Where did the stuff of that supernova come from? Did the star that exploded also condense from the debris of some previous explosion? If so, what formed the elements that went into making that star? And how long does our lineage go back? How many times have the elements now in our solar system been swallowed up by some star, digested, and regurgitated? A few times? Hundreds of times? Thousands?

This is the question that astronomer Alan Dressler and his colleagues at the Observatories of the Carnegie Institution, based in Pasadena, would like very much to answer.

"The history of chemicals in this galaxy is the history of chemicals that produced our bodies," said Dressler. "But the details of that aren't known."

That history is hard to sort out, in part because so much recycling has mixed everything up. "It's like adding salt and sugar to your coffee and then trying to separate them out," said Dressler's colleague, astronomer Wendy Freedman.

Different kinds of supernovae spew out different brews of elements. Each leaves its individual imprint, but together they sound like all the radio stations blaring at once—a whole lot of noise. "It's like a chorus of voices," said Dressler. "You have basses, you have tenors. But sometimes, the tenors can sound like basses, and vice versa."

And the mixing never stops.

"Things are constantly sloshing around in the galaxy," said Gus Oemler, director of the Carnegie observatories. Sometimes, smaller galaxies fall in, adding their ingredients to the brew.

To get a handle on the problem, astronomers go back to a simpler time—which in astronomy, of course, means looking out into space with high-powered telescopes. Specifically, astronomers are searching out relatively virgin stars—stars, that is, that have gone through very few recyclings. These stars stand out because they are only slightly contaminated with elements produced in previous stellar explosions. In other words, they are mostly just hydrogen and helium.

Alas, such stars are rare objects—only about one in every ten thousand. Enormous new twin telescopes going up now at Carnegie's site in Chile will soon join the search.

The 6.5-meter Magellan scopes will help the astronomers do what Oemler calls "archaeology on our own galaxy." Among other things, they will be looking for the traces of the very first stars, composed of the pure, unadulterated ingredients that ultimately got recycled into us.

It's an astonishing thought: to be able to look back to a time when nothing in the universe had been used before.

Not even the elements.

Simplicity

Science is a simple pleasure—the simpler, the better.

In fact, simplicity is often invoked as the gold standard of a sound theory. As Einstein put it: "Make everything as simple as possible, but not simpler."

There's a reason for this beyond sheer elegance: Simplicity implies that one has managed to see through the superficial complexity of a situation straight down to the bone, to distill its pure essence. Einstein was a master at this. What is more simple than the almost babyish equation $E = mc^2$? And yet, contained within those three terms are the fires that fuel stars, the radioactivity that melts rocks and moves the continents, and ultimately even the alchemy that turns sunlight into children.

Simplicity also suggests that a concept is well understood. Ongoing research is almost always confusing, but once the truth is grasped, the fog of complexity clears, and answers become transparent. It's easier to see where you've been than where you're going. Even for Sherlock, only in retrospect were things "elementary, my dear Watson."

A long legacy of astonishing simplifications leads physicists to suspect that truth is simple. Electric sparks and magnetic attraction, energy and matter, falling apples and shooting stars, all turned out to be but different aspects of the same things.

That's why present-day physicists scour for further simplifications still: Beneath the confusing collection of elementary particles and various forces, is there just one simple underlying entity?

And what about the still mysterious "dark matter" that appears to account for the vast majority of the universe? Ideas for its identity abound, but most require complex interactions among various forces and particles.

Fermilab cosmologist Rocky Kolb, however, has proposed that dark matter may be something far more simple—a consequence of vibrations in the vacuum of the newborn universe that got pulled apart by rapid expansion and became stable particles. (They would be hugely massive versions of theorized particles known as WIMPs, which is why Kolb calls them Wimpzillas.)

These dark matter particles would be produced by the same established mechanism that created the ripples in space that grew up to be clusters of galaxies. No need for anything fundamentally new.

"It's appealing to me because it's not something you invented to make dark matter," he said. It's appealing, in other words, for the same reason that Newton's linking of falling objects and orbiting bodies was appealing—it makes a complex situation simpler.

In truth, simplicity in physics is not so different from the economy we prize in other fields. Writing, for example. Writers of few words can often say more with less.

Princeton mathematician Ingrid Daubechies once told me that the best thing about becoming a MacArthur Foundation "genius" fellow was the chance to discuss her work with people far outside her field. She was especially pleased, in conversing with a poet, to realize that math and poetry boiled down to much the same exercise. You discover some essential truth, distill it to its pure form, and figure out how to communicate it to others.

Wise old sayings accomplish the same trick. "A stitch in time saves nine" is a lot more economical than: "Attack a problem as soon as it arises, because if you wait, you'll have a much bigger problem on your hands."

Or consider the rich balance between openness and skepticism contained in the admonition: "Always keep an open mind, but not so open that your brains fall out." How many pages of prose could say it better?

Of course, there's no guarantee that truth is simple. Many elegant, simple theories are wrong. And what's simple in one context may be horrendously complex in another. Fermat's Last Theorem could be stated in a sentence, but it took hundreds of pages and hundreds of years to prove.

Ironically, as the fundamental understanding becomes simpler, it often seems further removed from complex reality.

When my son was young, he liked to joke: "Atoms for dinner again, Mom?" Everything we eat is composed of protons, neutrons, and electrons—but it's a long way from those three simple particles to pizza and cake.

"There's a sense in which that's inevitable," MIT physicist Frank Wilczek told me. "The world doesn't change. It's still complicated. [But as explanations get simpler], you need to make longer chains of logic to make contact with the real world."

Einstein had much the same thought: "Although it sounds paradoxical, we could say: Modern physics is simpler than the old physics and seems, therefore, more difficult and obtuse."

Still, it's hard to deny simplicity's appeal. In a new collection, writer Arthur Clarke relates the story of some young newspaper reporters who once made a bet over who could write the best story in six words.

The winner was a fellow named Hemingway.

He wrote: "For sale. Baby shoes. Never used."

"Heartbreaking," says Clarke.

Simply amazing.

Complexity

Science seeks, above all things, a pure, elegant simplicity. Or so I've written many times before.

In fairness, I should confess that a chemist friend tells me I've been brainwashed by theoretical physicists.

True, the fundamental laws may be simple, he says, but everything interesting is complicated: the frenzied nerve firings that allow you to make sense of these words, or pick up a pen to write a nasty letter to the editor; the patchwork of continental plates whose scraping and shifting creates the ground on which you sit; the brew of chemical elixirs that churn up emotions; the dozens of molecules it takes to make up the aroma of chocolate. Rain, music, rocks, puppy dogs, sand dunes, fire—all unspeakably complicated things.

Even atoms, if truth be told, are complicated. We like to think of them as neat little planetary systems, electrons orbiting hard, sedentary centers. Nothing could be farther from the truth. At heart, they are a froth of restless uncertainty.

A single atom can't be red, sweet, hard, sticky, soft, shiny, loud, rude, or kind. It takes a village, or at least several million. And even then, it's how you arrange the atoms that matters. From dull carbon alone you can get soot and diamonds, graphite and soccer-ball–shaped Buckminsterfullerenes—depending on how you shuffle the pieces.

As for purity, it doesn't exist.

"If you were to look at the purest things in our environment," writes chemist Roald Hoffmann in *The Same and Not the Same,* "you would find that at the parts-per-million level, you might not want to know what is in there. Everything is in fact quite dirty."

The purest spring water, he tells us, "is a downright frightening mixture."

Even pure mathematical forms exist only in our imaginations. There is no such thing in this universe as a perfect square or circle. To the extent purity does exist, we probably wouldn't much care for it. Purity is boring. "Mountains are not pyramids and trees are not cones," says the teenage mathematical prodigy Thomasina in Tom Stoppard's play, *Arcadia.* "God must love gunnery and architecture if Euclid is his only geometer."

It takes many thousands of different kinds of molecules to make a person, almost as many to make a pear. Life is complicated in part because it has been pieced together by evolution, borrowing whatever worked from whatever ingredients were handy (off-the-shelf, as it were) at the time.

More intelligent design would no doubt have produced much more streamlined beings; a lot more like robots; a lot less endearing. The most beautiful people, after all, tend to be mixtures. (Think of Brazil.)

From complexity comes the capacity for difference. A sphere is a sphere is a sphere but no two snowflakes are exactly alike. And snowflakes are constructed only of simple water, a molecule of merely three atoms—H_2O.

Hemoglobin, by contrast, contains 2,954 atoms of carbon, 4,516 of hydrogen, 780 of nitrogen, 806 of oxygen, 12 of sulfur, and 4 of iron.

With so many varieties of each atom possible (including the

isotopes), and so many combinations of atoms in large biological molecules, it's probably safe to assume, writes Hoffmann, that "there are no two identical molecules in [a] Burmese cat."

Never mind two cats. There is no such thing as a simple individual.

To be sure, we like the idea that science can make things simple. It's part of the appeal. And so we get understandably angry when weather forecasters predict sun and then it rains; or scientists can't tell us exactly what's happening to the climate, or how to cure AIDS, or why people kill, or when a human life begins—or ends.

Even the physicists are pretty useless when it comes to keeping track of more than a few particles (or planets) at a time—at least not without high-powered computers to help them. Sedate as it seems, our nine-planet solar system will eventually descend into chaos simply because it has too many moving parts for long-term stability.

On the level of fundamental law, it's true that Nature seems to prefer simplicity. You can reduce a Burmese cat to quarks and electrons. But then you have neither Burmese nor cat.

Besides, we don't live at the level of fundamental law. On the human scale, things are messy; with so many different players and moving parts, it's impossible to keep track of what's going on, much less completely understand it.

So we shouldn't expect simple answers from science in a complicated world, whether they have to do with climate change or disease control, car safety or crime prevention.

"If we insist that [complex things] must be reducible, all that we do is put ourselves into a box," says Hoffmann.

And then, all we've reduced is ourselves.

Dazzle

P eople were lined up for at least a half a mile—staring, silent,
   transfixed. Kids on bikes stopped in their tracks, leaning
with elbows on handlebars. An old lady turned in her walker.
Fathers lifted toddlers on their shoulders. The Rollerbladers
braked. Dogs sat quiet on leashes, waiting for signs of life from
their suddenly still masters. Homeless people wrapped in blankets came to look, some trailing garbage. Even the runners
paused, putting aerobics on hold.

At first, I thought something horrible had happened. But,
no. It was only (only?) a sunset at Santa Monica beach. In a singular act of community, L.A. sun worshipers lived up to their
name, bidding Old Sol adieu with fitting reverence. Or, at least,
bonne nuit.

What is it about a sunset? Surely, it's not just the color. You
can get almost the same thing from a good box of paints.

Or even, if you like to play, by shining a flashlight through
water sprinkled with powdered milk. The milk particles, it turns
out, are just the right size to scatter the blue light preferentially,
so it leaks out the side, leaving only the red to travel the distance. It's exactly what air does to sunlight. So if you shine the
yellow flashlight beam through a glass loaf pan lengthwise, the
sides will look sky blue, but down at the end, you'll see a shim-
mering red-orange ball.

It's not that hard to cook up a sunset in a pan.

Okay. So maybe a loaf pan full of colored light isn't the

same as a whole sky aglow. But still, I think there's more to the way the sun moves us than mere (mere?) beauty.

At some level, I think even people who know nothing of physics understand just how remarkable a thing it is, this yellow star of ours. Turning matter into energy as surely as Rumpel-stiltskin turned straw into gold—two dozen ocean liners' worth of it every single second. Except that the sun's gold is really manna from heaven, the stuff we eat for breakfast, lunch, and dinner.

(The energy you use to read this sentence is powered, ulti-mately, by sunlight—perhaps first soaked up by some grass that got digested by a cow before it turned into the milk that made the cheese that topped the pizza. But sunlight, just the same.)

It's a godlike power, for sure. Perhaps that's why people were so upset when Galileo published a letter in 1613 suggest-ing that the sun was less than perfect; that it broke out sporad-ically with imperfections, like ordinary folk. The dark blotches Galileo saw suggested that the sun was up to something—not just some lazy orb, as the ancients had thought, with nothing to do but roll around heaven all day.

Now, we know the dark spots are relatively cool areas where the sun's magnetic fields get knotted up, twisted around by the star's rapidly spinning electrically charged atmosphere. It's a mystery, still, how these spots come and go, and the more as-tronomers learn, the more interesting the mystery seems to get.

To wit: The spots, which are associated with intense mag-netic activity, come and go in eleven-year cycles. As the last cycle's spots disappear near the equator, new ones pop up at the poles—but with magnetic poles reversed. No one can figure out how a star manages to turn itself inside out at such regular intervals, or why.

Or figure this: The entire sun pulses in regular five-minute intervals. (No wonder it's the whole planet's heartthrob.) In addition, it rings in dozens of different harmonics, which are now being studied by the equivalent of Earth-based stethoscopes placed strategically round the globe.

There's only so much you can learn from a distance, however, so NASA's sent up a spacecraft to bask in the sun's glow for a couple of years, collecting the star's exhalations on silicon detectors. In 2004, it's due to swoop down over Utah, where the samples will be collected, mid-air, by helicopter.

The mission, aptly named *Genesis,* is after no less than the story of our own origins. Some 4.5 billion years ago, the sun, planets, and moons all condensed out of the same maternal gas cloud; when the planets spun off, the sun held on to more than 95 percent of that primordial stuff. It still does.

Humanity really *is* a chip off the old block. If we can bottle the stuff and bring it home to our laboratories, we'll know a lot more about ourselves.

As the days get longer, I realize how much I've missed the sun during its long hibernation. How much I appreciate all it does for us: Putting tails on comets. Driving weather. Making pizza.

Finally. Here comes the sun. Goodness, gracious, great balls of fire.

Energy

If there was ever a pro at poking fun at the know-it-all pretensions of some physicists, it was Caltech's own late Richard Feynman. And more often than not, his ribbing struck some deep shard of truth.

Take his turn on that familiar everyday phenomenon: energy. We buy it, we save it, we waste it; we wolf it down in the form of jelly doughnuts, then desperately try to work it off at the gym; we soak it into our skin as sunlight and scream as it carries the roller coaster over the edge; it stirs breezes and powers stars; it's the quiet potential sitting in a tank of gas and the foamy violence of an ocean wave; it's coiled into springs and frozen into matter. If there is any justice in the world, some Enron execs will go to jail over it.

And yet, as Feynman points out, "It is important to realize that in physics today, we have no knowledge of what energy *is*."

It's even worse than that: We don't even have a sensible way to name it or measure it. "Physicists sometimes feel so superior and smart that other people would like to catch them out once on something," Feynman said. "For those who want proof that physicists are human, the proof is the idiocy of all the different units which they use for measuring energy."

There are calories and joules and ergs and electron volts and BTUs and kilowatt hours, to name a few—all measuring this same nameless thing.

But here's the absolutely amazing thing about energy: The

amount of it in the universe is absolutely constant. It never changes. It only alters its form.

Consider the energy you used to get up this morning, open the paper, and read the news. It came, of course, from that chicken sandwich you ate last night. The chicken got its energy, in turn, from the corn, which got it from the sun.

And that minor earthquake last week. The energy for that bubbled up from deep inside the earth where radioactive atoms melt rock, moving continents around. Those atoms in turn got crammed full of energy during the gravitational collapse of some long-gone star, later to be spewed into space; some of the debris condensed into the glob of matter we now call Earth. The gravitational energy that crushed the star, in turn, can be traced back to the Big Bang, the explosive expansion of the universe.

It's an endless regression of recycling.

And it doesn't stop there.

The universe got the energy to bang in the first place from the energy of the vacuum of empty space—which, oddly enough, is also packed with energy. This energy of emptiness comes from the perpetual quivering inherent in everything thanks to the innate uncertainty of quantum mechanics. A large enough quiver, according to some cosmologists, could have begun the whole shebang.

This sobering thought led MIT cosmologist Alan Guth to his now infamous conclusion: The universe is the ultimate free lunch. "Conceivably," says Guth, "*everything* can be created from nothing."

Of course, the "nothing" that we're talking about here is really a very energetic speck of vacuum. But as far as the universe is concerned, the energy of a vacuum and the energy of a Mars bar are substantially the same. Energy is energy. Whatever that may be.

"As far as we know, there are no real units, no little ball bearings," said Feynman. "It is abstract, purely mathematical, that there is a number such that whenever you calculate it, it does not change. I cannot interpret it any better than that."

Of course, if the amount of energy in the universe never changes, you might well wonder how can there be such a thing as an energy problem. Why the Mideast? Why Enron? Why world hunger, for that matter? Why not, to paraphrase the queen, "Let them eat vacuum"?

The reason is that while we don't ever use up energy, we do use up useful energy. Hiroshima before and after the atomic bomb contained, roughly speaking, the same amounts of energy, but one would hardly call them equivalent. A tree spends centuries soaking up the energy of the sun and soil, and releases it all in an hour in a fire. It can take a thousand years to turn muscle and thought into cities, eons to pack the energy of sunlight into fossil fuels, millions of years of steady gravitational pressure to light up a star. Using energy is often quick and easy, but making it useful is long and hard.

Conservation of energy *is* built into nature, as Feynman points out. "But she does not really care; she spends a lot of it in all directions."

To spend it wisely, now that is up to us.

The Real World

As I've written before, the late physicist Frank Oppenheimer used to get furious when people told him he had to behave in a certain (usually conventional) way because this was, after all, the "real world"; he would simply have to adjust.

Frank never thought so. Instead, he insisted: "It's *not* the real world. It's a world we made up. We can make it another way."

While Oppenheimer was usually referring to societal norms or the administration of the science museum he founded, he was also speaking as a true physicist.

After all, telling the difference between the "real world" and the "world we made up" is a major preoccupation for physicists. Are black holes real? What about exotic, as yet undiscovered, particles of "dark matter"? What about the repulsive energy of empty space? What about time?

One needn't get so exotic. Consider this book. Is it real? Certainly, it's made of chemical compounds, which are made of molecules, which in turn are made of atoms. Are atoms real? If you stood inside one, you'd see enormous amounts of empty space, with little fluffs of electric charge buzzing around the far periphery; in the center, a tiny nucleus composed of protons and neutrons.

Are protons real? Anything as small as a proton behaves more like a wave than a particle. Not a wave of matter, mind you, or even a wave of energy, but a "probability wave." This probability wave tells you the likelihood of finding the proton in a given place.

Under close-enough scrutiny, the real world of matter as we know it evaporates. We're left with patterns and relationships. It's the pattern of waves that gives the proton its substance; the relationship between protons and gravity and electricity that makes the world go round. What's solid and unchanging are these patterns and their interactions.

In particular, what lies beneath the illusion we call "reality" are symmetrical relationships that look the same no matter how you turn them. "In the physicist's recipe for the world, the list of ingredients no longer includes particles," notes Nobel laureate Steven Weinberg of the University of Texas. "Matter thus loses its central role in physics: All that is left are principles of symmetry."

Einstein proposed that we believe things are "real" only insofar as we relate to them through our senses of sight, hearing, and touch and instruments that are extensions of our senses. "The concept of the 'real external world' of everyday thinking rests exclusively on sense impressions," he wrote.

But sense impressions are processed in the mind—itself a slippery concept.

"So what is this mind, what are these atoms with consciousness?" asked Richard Feynman. "Last week's potatoes! That is how we can remember what was going on in my mind a year ago—a mind which has long ago been replaced."

Feynman points out that no single atom in the brain is a permanent resident. "The atoms come into my brain, dance a dance, then go out: always new atoms but always doing the same dance, remembering what the dance was yesterday."

If pinning down reality is a matter of seeing consistent patterns, then humans are well equipped, because ferreting out patterns is what we do best: We see patterns in the stars, on the moon, in the cracks in the ceiling, in the orbits of the planets.

Arranging things into patterns makes them easier to understand. But only sometimes do the patterns point out "real" relationships.

The orbits of the planets are connected by gravity, for example, but the stars in the Big Dipper are connected only by our imagination. Both are real, but the motions of the planets are manifestations of natural law; the dipper in the sky reflects human culture projected on nature. All too often, we mistake the "real world" for one that exists mainly in the human brain.

"Having lost the gods, we fall in love with the beautiful idols we can raise in their places," writes Amherst College physicist Arthur Zajonc in his book, *Catching the Light.* "Atoms, quarks, tiny black holes . . . they are reified, garlanded, and dragged forward to assume a place in the temple. Calling them real, we animate them. . . ."

In the end, finding out what's real may require redefining what we mean by reality. After all, science often requires that we go beyond sense impressions (as well as common sense). No matter how real or unreal something might seem, it ultimately has to stand up to experimental scrutiny and theoretical consistency.

The chair is real because you can sit on it. Newton's laws of gravity are real because they keep the planets in orbit around the sun. Real enough to command our attention, in any event. Real enough so that they can't be easily ignored. Or as Weinberg puts it: "When we say that a thing is real, we are simply expressing a sort of respect."

Magnetism

The sun has a flair for putting on spectacular shows, and the one in the Spring of 2001 was a real doozy. Images taken at various National Science Foundation–funded observatories caught our local star in the act of spitting out glowing filaments twenty-two times the diameter of the Earth—dragon's breath streaming into space for a hundred thousand miles.

Such flares create the beauteous aurora borealis, or northern lights, which in this case reached as far south as Palm Springs. They can also scramble ground communications and puff up the atmosphere enough to drag satellites to early deaths.

Oddly, these seemingly special effects are created by one of the most familiar (if least understood) forces in the universe, the same invisible influence that keeps a refrigerator magnet stuck to the door. And increasingly, good old magnetic fields are emerging as major players in the universe at large—sculpting everything from the last gasps of dying stars to the dynamic centers of galaxies.

Astrophysicists shudder at this turn of events, because magnetic fields are among the most complicated phenomena they have to deal with. "They're what every astrophysicist loves to ignore because they complicate things enormously," says UCLA's Mark Morris.

No one understands completely how the sun slings out the streams of electrically charged particles produced in the flares—

although it has something to do with how the tangled magnetic fields inside the star get twisted as it spins, then snake to the surface and snap. Still, the sun's act is but a sideshow to what magnetic fields are up to in the rest of the universe.

A few years ago, for example, astronomers announced that they had discovered surprisingly strong magnetic fields hanging out in the seemingly empty spaces between galaxies—a real puzzle. A magnetic field is produced by moving electric charges, so how do you create strong fields where you don't have many particles?

Other astronomers recently figured out that magnetic fields are responsible for molding multicolored clouds of glowing gas blown off like smoke rings from dying stars.

Still other astronomers reported that the strange asymmetrical jets blown out of Supernova 1987A were probably caused by—you guessed it—magnetic fields, produced inside the collapsing star.

Magnetic fields in space are a mess to understand because they combine some of the most complicated phenomena in physics: magnetism and fluid flow. The fields are created by flows of electrically charged particles that can swirl around unpredictably like eddies in a stream.

In turn, the magnetic fields can steer the flow of particles, directing them this way or that. This creates more fields, and so on and so forth, the fields and flowing particles pulling each other up by their sometimes beautiful bootstraps.

It's not surprising that astrophysicists would just as soon not get pulled into this vortex. "I deal with magnetic fields all the time, but there's a lot of things I don't understand about them," said the University of Rochester's Adam Frank, whose work helped solve the mystery of the multicolored clouds.

Yet the phenomenon seems unavoidable.

For example, some years ago Morris and colleagues found strange magnetic filaments running hundreds of light-years right through the center of our galaxy, in uncannily straight and narrow rows. Dozens of these tightly bunched parallel fibers have been found, all running perpendicular to the galaxy's nearly flat disk.

The only reasonable explanation seems to be very strong magnetic fields. But if so, where did they come from? Probably they are primordial magnetic fields that became concentrated in the center of our galaxy when it formed, Morris said. But that only pushes the question back in time. "Nobody has figured out how magnetic fields first came into being in our universe," he said.

In fact, all the magnetic fields writhing inside stars and interstellar clouds can probably be traced back to the early universe, Morris said.

Even on small scales, of course, magnetism seems magical. A refrigerator magnet sticks to the door because countless unseen atoms inside line up in artfully arranged rows; each atom spins up its own magnetic field with the help of its orbiting electrons.

Magnets manage to speak to each other across the chasm of empty space. Opposite poles pull on each other like invisible hands. Likes shove each other away with a palpable push. "It's freaky," said Frank, "because you can feel the force, but you can't see it."

And it's not just mystical; it's useful.

Doctors read the magnetic fields of atoms spinning inside our bodies to create images of what ails us. The spinning iron core of the earth creates a magnetic field strong enough to shield us from those streams of particles the sun spits out in our direction—a global magnetic umbrella. Flipping magnetic fields even store the information kept in computers.

Undeniably attractive, magnetism steers the compass of the universe. Wherever you go, it seems, this force is with you.

Sand Castles

It's odd how the everyday things are often the most mysterious. How churning mind springs from sedate matter, for instance. Or the origin of space and time.

Which came first, the chicken or the egg? What is matter, and why does it have mass? How exactly does gravity work? Scientists still don't know what drives the erratic magnetic compass of the Earth, or where the moon came from, or why the sun changes its spots in cycles of eleven years. They don't understand how proteins knot themselves into shape, or why snowflakes crystallize into ever-changing patterns.

So while some scientists explore the exotic frontiers of the universe, hunting for wild specimens such as quasars, pulsars, or quarks, others have learned what Dorothy discovered when she returned from Oz: Sometimes the richest treasures are buried in one's own backyard.

These days, physicists at the forefront of science can be found putting enormous energy into understanding something that finds its way into the backyards of many Southern Californians—sand.

Sand may seem like child's play, but for physicists, it's frustratingly hard to get a handle on. The problem is, it's weird stuff. It's made of solid grains, but it flows like a liquid. Dry, it floats around on the wind like dust, but wet, it can be sculpted into castles.

Actually, sand forms its own special category—not solid,

not liquid, not gas, but "granular material." Studying the behavior of sand tells scientists a great deal about the behavior of any pile of stuff—whether it's piles of magazines, peas, soil, salt, or aspirin. These granular materials exhibit bizarre behavior with no obvious explanation. For example, if you shake up a jar filled with granular material of different sizes, the biggest grains always rise to the top. That's why the Brazil nuts always wind up on top of the can of mixed nuts, and the peanuts on the bottom.

Somehow, granular materials manage to sort themselves by size.

This odd property causes a major problem for many industries, including makers of pharmaceuticals, who often have to mix powders together, and need smooth, evenly blended concoctions.

More challenging to physicists, there's no good explanation why inanimate matter should organize itself so readily. Over time, things normally mix, not unmix. If you leave a perfume bottle in the corner of the room, sooner or later all the perfume molecules will waft out, wandering willy-nilly about the house. If you leave the refrigerator door open, pretty soon the cold air will mix with the warm air of the room, and both room and refrigerator will be the same temperature.

But granular materials somehow spontaneously line up in layers. They appear to disobey the laws of nature. The question is, how?

Physicists also don't understand exactly how a sandpile stands up. You would think that a mound of sand would feel the most pressure at the point where the pile is highest: in the middle. Surprisingly, the maximum pressure point is actually a ring that forms around the apex.

This might explain why children can build tunnels through

sand castles without causing them to crash; somehow, invisible arches inside the castle must be supporting the load. Susan Coppersmith of the University of Chicago proposed in the journal *Science* that sand somehow distributes its weight unevenly, forming these "stringy" lines of force. More recently, an article in the journal *Nature* explored the question of why wet sand behaves so differently from dry sand. Apparently, the water acts as an adhesive that makes the grains stick together, but the effect is surprisingly large.

"Small quantities of wetting liquid can thus dramatically change the properties of granular media," concluded physicist Peter Schiffer of the University of Notre Dame.

Physicist Per Bak of Brookhaven National Lab, meanwhile, has developed a new branch of science built on sand. Called "self-organizing criticality," it explains how avalanches are triggered in sandpiles, and how the pile retains, over time, the same general shape. Bak thinks the theory will also help to explain everything from earthquakes to stock market crashes.

Perhaps the poet was right: You can find a whole world in a grain of sand.

Resistance

My friend the physicist used to get very angry with people who wouldn't argue with him. He thought you couldn't make any progress pushing the boundaries of understanding if other people didn't push back. He knew that pushing back was an essential part of pushing onward.

As a driving force of nature, pushing back is often vastly underrated. People wish their lives could run smoothly without resistance, just as engineers wax enthusiastic about superconducting power lines that carry electricity friction-free.

We forget that, without resistance, life would be unlivable. And it's not just a matter of saying no to drugs or laying down the law for rebellious teenagers or two-year-olds. Without resistance, we wouldn't be able to dance, play golf, or even blow-dry our hair. We think of resistance as an impediment, when in fact it's an enabler.

Consider the simple act of walking. If you've ever slipped on a banana peel or skidded on an icy patch of road, you know the hazards involved in having your usual source of resistance suddenly removed. You can walk forward only because the street pushes back. You might think this is not rocket science, but, in fact, it is. A rocket can race forward only because something—if only hot air—is racing backward with an equal and opposite force.

For that matter, it's pushing back that allows you to push on anything. Pound the desk with your fist and it hurts because the desk pushes back; pound a marshmallow and nothing much happens, because marshmallows can't push back. In fact, it is physically impossible to hit anything (or anyone) harder than it can hit you—which is a good thing to remember the next time you're deciding whether to punch a wall or a pillow.

Without resistance, the world as we know it wouldn't exist: It's resistance that makes toasters and hair dryers hot, that powers the recoil of the golf ball from the club, that propels the ballerina in her multiple pirouettes.

And that's only the beginning. Without a pervasive pushing back, we wouldn't have matter at all. Normally, we think of matter as that which has mass, and mass is defined by inertia. The harder something is to push around—that is, the more it pushes back—the more massive it is.

Where does this resistance to motion come from? Why are sofas harder to push across the floor than shoes?

Physicists believe that particles pick up mass as they slog their way through an invisible cosmic molasses known as the Higgs field. If it weren't for the resistance of this unseen stuff, everything would travel at the speed of light and particles of matter would never slow down; they'd never have the heft to clump into atoms or anything else.

So, resistance is useful as well as inevitable. It's also a lot harder than most people think. We often equate pushing back with doing nothing—like a wall that just sits there as a car crashes into it. Like the ease implied in the expression: "Just say no."

Nothing could be farther from the truth. In terms of the energy required, there is no difference between accelerating and decelerating, a start up and a stop down.

This is a hard fact of nature that NASA scientists face continually. Say they want to land a spacecraft on Mars. Obviously, they need enough fuel to fight Earth's gravity and blast the rocket into space and on to Mars. But if they want a soft landing, they also need to carry enough fuel to slow the spacecraft down before it reaches the planet's surface.

This is a good thing to remember the next time you're struggling to quit a bad habit. Whatever energy you put into creating the habit is the amount of energy you will need to push it out the door.

Pushing onward and pushing back are two sides of a coin. One without the other is as unthinkable as one hand clapping.

Or as Paul Hewitt, author of *Conceptual Physics,* often puts it: "You cannot touch someone without being touched in return."

Happenstance

As much as physicists hate to admit it, not everything in nature happens as a result of natural law. Strange to say, much of the universe is an accident.

Take the family of planets, for example. Johannes Kepler spent decades trying to figure out the harmonics behind the orbits of Earth and its neighbors. He tried to make them trace perfect, resonant patterns that would sing out the music of the spheres.

He failed because the planets didn't fall into place according to plan. They just fell. When the solar system condensed from its primordial gas cloud, clumps of matter appeared here and there, by chance. One of them gave rise to us.

"We now understand that the planets and their orbits are the results of a sequence of historical accidents," writes physicist Steven Weinberg in *Dreams of a Final Theory*.

He concludes: "The most extreme hope for science is that we will be able to trace the explanation of all natural phenomena to final laws and historical accidents."

It's often hard to tell which phenomena follow from law and which happen by happenstance, however. While Kepler spun his mental wheels trying to find reason in chance, other scientists dismissed important clues as "mere" accident. Geologists wrote off the almost perfect fit between the coasts of South America and West Africa as a funny coincidence; physicists dismissed the uncanny equivalence of inertia and gravity as the same. The former turned out to be the key clue leading to the discovery of continental drift; the latter to the warping of space-time, with its attendant black holes, and the rest.

The distinction between law-abiding and accidental connections is crucial in science in part because phenomena that follow laws can be predicted, while accidental ones cannot. And it's certainly one reason physicists would very much like to know whether or not the laws of physics themselves are an accident. Or as Einstein put the question: Did God have a choice?

If the laws of physics are set in stone—if there's only one possible mass for an electron or four possible dimensions for space-time or one possible strength for gravity—then God did not have a choice. Knowing the laws of physics would allow us to predict the universe as we know it. Yet recent thinking suggests a more unsettling scenario: The universe could have evolved in many different ways out of the primordial chaos—all of them equally plausible. Just as ice can crystallize from water in just about any form, so the matter and energy in the universe could have crystallized in any number of ways.

The fact that things fell into place in a way that allows life to exist, according to this argument, is simply serendipity. Life got lucky. On the other hand, maybe the universe we perceive is the way it is because it's the only kind of universe we *can* perceive. Other kinds of universes may well be out there. But their physical laws would make life impossible.

Curiously, the evolution of life itself is ruled by a strange marriage of cause and chance. The genetic alterations that can lead to changes in species are random events—stray cosmic rays zapping atomic bonds.

But biological imperatives ultimately determine success or failure. While chance plays a role, suitability to habitat plays a larger one.

Physicist Lee Smolin of Pennsylvania State University suggests that the universe itself evolved by a kind of natural selection. "The laws of nature themselves, like the biological species, may not be eternal categories," he writes, "but rather the creations of natural processes occurring in time." Roughly, according to Smolin, our "accidental" universe evolved like this:

First, he assumes that multiple universes are continually sprouting into existence from the back sides of black holes. (In essence, space-time gets pinched out of existence inside a black hole, but gets reborn out the other end into a new baby universe.) So universes that create many black holes give rise to more baby universes than universes without black holes.

Black holes are created by stars that explode and collapse under the weight of their own gravity. So universes with lots of stars are more "fit" to survive than others. As it turns out, the same properties that produce stars make the universe a fit place for intelligent life-forms.

So partly by plan, partly by accident, our universe (and perhaps others) evolved to produce beings like us. As Smolin explains it, the universe is more like a city than a clock. A clock needs a clockmaker to create it and to wind it up. But cities—like species—spring into existence seemingly of their own accord.

"If a city can make itself, without a maker," he concludes, "why can the same not be true of the universe?"

The good news in all this is that the universe is not, as some people suspect, an accident waiting to happen. It's an accident that happened 13 billion years ago.

Ghosts

With perhaps unwarranted exuberance, physicists at Brookhaven National Laboratory announced possible glimpses of a new branch of the particle family tree—a whole new lineage of previously unknown relatives with weird-sounding names like squark, selectron, gluino.

Curiously, these particles left their tracks not in mud or photographic plates, but rather by the effects they had on, well, nothing. Or rather, what most people think of as nothing, which is the vacuum of empty space.

Not that physicists were surprised to find something going on in nothing-at-all. Since at least the 1930s, they've known empty space is the site of some of the strangest activity in the universe—a stage where ghostly "virtual" particles pop in and out of existence like bubbles in champagne. The very real effects of these virtual goings-on in the vacuum have by now been proven experimentally to almost a dozen decimal places.

How does the vacuum make itself seen? Well, for one thing, it doesn't let elementary particles run around naked. The vacuum makes particles as modest as brides in veils; they never appear without chaperons in the form of those ghostly "virtual" particles. This veil obscures, for example, the particle's true electric charge. The naked charge is actually substantially larger than the shielded one.

"An electron sitting quiescent in empty space is not quiescent at all," physicist John Wheeler describes the situation. "As we zoom in on it with hypothetical microscopes of higher and higher power, we see a more and more lively neighborhood around the electron.... And the closer we get, the more violent the activity becomes. The 'isolated' electron is the nub of a seething volcano."

The Brookhaven physicists didn't look at electrons, but rather at their heavier cousins, called muons. And they weren't measuring electric charge, but something called magnetic moment, a kind of intrinsic magnetism caused by the muon's spin.

Physicists already knew, of course, that the vacuum was distorting their measurements of the muon, and they had calculated the effect with great precision. They can do this because the virtual particles that bubble out of the vacuum are ghosts of normal particles—identical in every respect, except that they don't quite stick around long enough or have enough energy to become real. Imagine them as subatomic Cheshire cats.

Each known species of particle has its own virtual doppelgänger. Imagine them as Cheshire dogs, goats, and zebras, popping in and out of the vacuum. If you add up all their comings and goings, you can predict exactly how the vacuum is going to affect the spinning muon, or anything else.

And until recently, the calculated effect and the effect measured in experiments matched exactly.

However, Brookhaven's unprecedentedly precise measurements suggest that there's more to the vacuum than previously met the experimenters' eyes. The entire known virtual particle zoo isn't enough to account for the effect they see. And one possible explanation is that the previously unseen relatives have finally been coaxed out of the closet—virtual unicorns and leprechauns, so to speak. (Only their real names would be charginos and smuons and neutralinos.)

Previously unknown particle ghosts aren't the only explanation for the experimental results, of course. The measurements could herald something even more exotic—like the first glimpse of extra dimensions. Or far more likely—like a statistical fluctuation. Or even a mistake.

Either way, the goings-on in the vacuum won't go away. Hidden in its depths are some of the thorniest problems in physics.

Indeed, the very idea that such strangely monikered characters as charginos and smuons exist at all came from attempts to solve just such a "vacuous" problem. Specifically, all that activity in empty space imbues it with enormous energy. And since energy has weight according to $E = mc^2$, the measurable effect on the universe should be enormous as well.

Alas, it's hardly (if at all) even seen.

One explanation is that every known particle (such as the electron) has a previously unknown partner (such as the selectron) that cancels its effect on the vacuum. This family of partner particles—officially called supersymmetric particles—are precisely what might have bubbled up at Brookhaven.

Whether it's time to break out the champagne and welcome the newcomers into the family fold is not yet clear. Further experiments (or better calculations) may send them back into hiding—if they exist at all.*

But the Buddhists are certainly on to something when they say that nothing holds the key to everything. Philosophers, too. As Sartre famously put it: "Nothingness haunts being."

Which only goes to show that Buddhists, philosophers, and physicists have, at the very least, a lot of Nothing in common.

Or in the words of Stanford physicist Leonard Susskind: "Anybody who knows all about nothing knows everything."

---

*As it turned out, an error in a complex calculation may well have made the ghosts go away. See "Blindsighted."

Symmetry

Even in sunny Los Angeles, it's probably fair to say that everybody loves a snowflake. But beyond the sheer pleasure of catching one on the tip of one's tongue, snowflakes teach lessons about truth and beauty—surely among the oddest couples to share accommodations in the house of science.

What does truth have to do with beauty, and how can snowflakes enlighten their relationship?

Snowflakes are beautiful because they embody just the right amount of symmetry. And for whatever reason, people find symmetry beautiful—whether in the mathematically perfect spirals of snail shells, the harmonies in music, the well-ordered arrangements of diamond crystals, or the multifaceted reflections in kaleidoscopes or decorative tiles.

In physics, symmetry has an uncanny ability to lead scientists to the truth. Recently summing up fifty years of progress in fundamental physics, the director of UC Santa Barbara's Institute for Theoretical Physics, David Gross, concluded: "The secret of nature is symmetry."

He advised his colleagues: "When searching for new and more fundamental laws of nature, we should search for new symmetries."

If you don't think physicists take symmetry seriously, consider that they are spending billions of dollars to track it down. The most conspicuous consumer is the Large Hadron Collider, currently being constructed by a collaboration of dozens of nations at the world's foremost physics lab, the European Center for Particle Physics.

When completed sometime around 2008, the collider will accelerate protons to near the speed of light, then smash them together in miniature versions of the Big Bang. What researchers hope to find is symmetry.

Symmetry has a slightly different meaning to physicists and mathematicians than it does to laypeople. Something is symmetrical in the "technical sense" to the extent that you can change it and the change doesn't make a difference. It doesn't make a difference if you make something bigger or smaller, turn it upside down or inside out, or put it on backward.

Snowflakes are somewhat symmetrical in that you can't tell the difference between a snowflake and its mirror image, or a snowflake turned upside down. But if you rotate the snowflake slightly, it looks different; so the snowflake isn't perfectly symmetrical. A circle is far more symmetrical: It doesn't change no matter how you rotate it.

You also can't melt a snowflake and expect it to look the same as it did before. In fact, although it's rare that two snowflakes look exactly alike, melted snowflakes almost always do, because any two drops of water tend to look alike. From the physicist's standpoint, the water that snow melts into is far more symmetrical than (though perhaps not as beautiful as) the snowflakes it came from.

In effect, the physicists at the Large Hadron Collider will be melting matter to find the perfect symmetry hidden underneath. At the extremely high temperatures they hope to achieve, the various particles of nature will "melt" into a more homogeneous state, just as distinctive snowflakes melt into indistinguishable puddles of water.

Why would they want to do that? To find out what fundamental features of the universe do not change—which hidden
symmetries lie behind the apparent diversity they see.

They would like to know, for example, why the quarks that
make up protons have a lot more mass than electrons. Why is
there a difference—a lack of symmetry—between the two? At
very high temperatures, they hope to re-create the symmetry
that existed before particles acquired different masses, a process which somehow "broke" or destroyed the symmetry.

At the same time, they hope to find out why the particle
world as we know it contains two very different kinds of particles: those you can put your hand through (such as light particles, or photons), and those you can't (such as those that make
up atoms). Call them apples and oranges.

At the super-high energies of the Large Hadron Collider,
however, physicists believe they will find a hidden symmetry—
a supersymmetry, or "SUSY," as it's affectionately called. According to SUSY theory, each particle teams up with a counter-
part in the other particle family to make a perfectly symmetric
team. Quarks, which are matter particles, are paired with light-
like particles known as squarks. The supersymmetrical "spart-
ner" of the photon is the photino.

Supersymmetry should reveal how the apples and oranges
that seem so different on the surface fell off the same primordial tree.

Meanwhile, a new Stanford particle accelerator called the
"B-Factory" (because it explores the behavior of particles called
B-mesons) is designed to explore the *absence* of symmetry. If
anything, this is more important (to us, at least) than symmetry.

Without symmetry breaking, we wouldn't be here. The newborn universe gave birth to matter and antimatter in equal—

that is to say, symmetrical—amounts. Now, we have only matter. Somehow, when the universe "froze" into its current state, it lost some of its symmetry.

The universe, in other words, is a lot like a drop of water that froze into a snowflake.

Imperfection

We live in an imperfect universe. This comes as news to no one.

What may surprise people, however, is that our universe exists only because of its imperfections. In fact, when people say "nothing is perfect," they are literally correct. Nothingness—and only nothingness—is perfect. Everything else is a little bit off.

Consider the very early universe—a state of pure, perfect nothingness; a formless fog of undifferentiated stuff: featureless, uniform, pure.

Perfection actually can be well-defined in physics by the idea of "perfect symmetry." It means that no matter how you try to change something, it doesn't make a difference. Look left or right, on the large scale or small, move fast or slowly, turn it upside down; it doesn't make a difference.

This is the perfect sameness we hear in utter silence, or see inside a cloud. It has no signposts, no direction, no flaws—nothing at all to make any one piece of it different from the rest.

"We have, in our minds, a tendency to accept symmetry as some kind of perfection," wrote the late physicist Richard Feynman, in his *Lectures on Physics.*

The closer physicists get to understanding the fundamental laws that rule the universe, the closer to perfect symmetry they come. And yet, our universe is far from this perfect state of grace: Forces are different from particles; electrons are different from quarks; gravity is different from electricity; and matter is different from antimatter.

"The reality we observe in our laboratories is only an imperfect reflection of a deeper, more beautiful reality," writes physicist Steven Weinberg. Physicists like Weinberg are in search of an ultimate theory of physics that displays "all the symmetries" of this lost perfection.

What shattered this primordial perfection?

One likely culprit is the so-called Higgs field, the subject of an international search. If it exists, the Higgs field literally took this formless perfection and froze structure into it, the way freezing imparts crystalline structure to amorphous water.

Water is perfectly symmetrical, but ice is not. Moving up is not the same as moving sideways. Freezing destroys the sameness.

Physicist Leon Lederman compares the way the Higgs operates to the biblical story of Babel. The citizens of Babel, you may remember, all spoke the same language. When they tried to build a tower up to heaven, however, God got mad and confused their speech—so they couldn't communicate with each other.

Like God, says Lederman, the Higgs differentiated the perfect sameness, confusing everyone (physicists included).

If true, this idea has wide-ranging implications. Normally, the Higgs is invoked only to explain how particles have different masses—why a quark is heavier than an electron, for example.

But influence of the Higgs (or the influence of something like it) could reach much farther.

In fact, something like the Higgs may be behind many other unexplained "broken symmetries" in the universe as well. For example, why is electricity so different from gravity? Why is our universe made of matter but not antimatter—even though the two appear to be created in precisely equal amounts? If there are really ten dimensions of space—as popular theories suggest—why are only three large enough for us to perceive?

The Higgs, says Fermilab physicist Joe Lykken, "potentially does a lot."

In fact, something very much like the Higgs may have been behind the collapse of the symmetry that led to the Big Bang, which created the universe. When the forces of nature first began to separate from their primordial sameness—taking on the distinct characters they have today—they released energy in the same way as water releases energy when it turns to ice. Except in this case, the freezing packed enough energy to blow up the universe.

Feynman wondered why the universe we live in was so obviously askew. "No one has any idea why," he wrote.

Perhaps, he speculated, total perfection would have been unacceptable to God. And so, just as God shattered the perfection of Babel, "God made the laws only nearly symmetrical so that we should not be jealous of his perfection."

However it happened, the moral is clear: Only when the perfection shatters can everything else be born. In the end, we owe everything to imperfection.

Numbers

What could be more dry than a statistic? More indifferent than a number?

To be treated like a number, in common parlance, is to become an entirely replaceable part—an object lesson in depersonalization.

It makes you wonder what physicist Richard Friedberg could have had in mind when he wrote in *An Adventurer's Guide to Number Theory*:

> Two is solid and tingly, like the Liberty Bell.... Eight is rough and hard like a stone, and 10 is smooth like a pebble on the beach. Nine...seems ready not only to ring but to shatter and burst like a fruit.

Friedberg's numerical affections may come as a surprise to laypeople, but not to mathematicians, who know there's more to numbers than simple counting. Numbers don't just line up like well-ordered rows of dominoes, keeping track of quantities. They inspire love, hate, fear, amusement, friendship.

Sometimes, numbers are to die for.

Consider, for example, the oft-told tale of the discovery of so-called irrational numbers. Irrational numbers cannot be expressed as the ratio of two whole numbers—which is to say, in some sense, that they can't be expressed exactly at all. Numbers like pi (3.14159—ad infinitum) and the square root of 2 (1.41421—ad infinitum) just trail on endlessly.

Such fuzzy indefiniteness didn't sit well with Pythagoras, the ancient Greek geometer who made a religion out of the seeming perfection of numbers. Pythagoras was so upset by the discovery of this obvious flaw, the story goes, that his disciples were pledged to secrecy. A rebel intent on taking the word out to the larger world was drowned at sea (by the gods, or his jealous colleagues—depending on who's telling the tale).

The discovery (or invention, if you will) of the number 0 was greeted with similar horror. Zero was regarded as the creation of the devil, and for a time, edicts were issued in Florence that forbade the use of the number and the new "Arabic" system that ushered it in.

"And, as usual, prohibition did not succeed in abolishing, but merely served to spread bootlegging," writes Tobias Dantzig in *Number: The Language of Science*. He notes that despite the edict, thirteenth-century merchants continued to use zero and its fellow numerals as a kind of secret code.

Others merely scoffed at zero, dismissing the new digit as a puffed-up pretender to the status of number. Sniffed one unidentified fifteenth-century source: "Just as the rag doll wanted to be an eagle, the donkey a lion and the monkey a queen, the cifra [zero] put on airs and pretended to be a digit."

And zero's alter ego, infinity, has been irritating mathematicians and physicists ever since it was discovered. Infinity, being without end, was widely regarded as a direct encroachment on holy territory. "The last number," writes Dantzig, "belonged to God."

Some numbers even had what amounted to moral qualities. The number 1 was reasonable; 2 opinionated. Followers of Pythagoras prayed to the number 4 (which stood for justice).

These days, mathematicians don't pray to numbers, but they do put them on the couch to analyze their personalities. "The number theorist strikes up a closer acquaintance and soon learns intimate details," writes Friedberg, including "likes and dislikes." Numbers, like people, can be complex, perfect, imaginary, amicable, surreal, transcendental, excessive, square, prime.

A perfect number, for example, is one that is equal to the sum of its divisors—like 6. (Six can be divided by 1, 2, and 3, and 1 plus 2 plus 3 add up to 6.) Excessive numbers, like 14, add up to more than the sum of their divisors. Amicable numbers are pairs in which each is the sum of the divisors of the other, like 220 and 284. The number 64 is both a square (of 8) and a cube (of 4).

These are things you might never know, Friedberg points out, if you merely regard numbers as a kind of alphabet for counting. In common arithmetic, he laments, "Sixty-four is not allowed to flaunt its special traits but must keep step, in front of 65 and after 63."

So forget all that mystical stuff about the number 666 belonging to the devil. Next time you want to know about transcendental qualities of numbers, don't go to a psychic. Ask a number theorist. And don't worry if you disagree with Friedman's characterization of the number 7 as "dark and full of liquid, like oil when it oozes from the ground." Or 5 as "pale but round like a ball."

The good mathematician leaves room for personal interpretation. "Perhaps you see 7 as a pincushion," he says, "and 5 as a bright spot of light."

Symbols

Mathematics has a strange gift. It is "unreasonably effective" in places where it has no right to be. Like a hammer that turns out to be good at scrambling eggs, or a basketball star who decides to become a rapper. Math plays a starring role not only in fields where you might expect it—say, physics—but also in such unlikely places as Hollywood filmmaking, ecology, medicine, traffic control.

What's a nice equation like you doing in a place like that?

In the summer of 2000, the superstars of mathematics gathered at UCLA to celebrate this strange flowering, as well as to look for fertile ground to plant new fruit. It was a gathering the likes of which had not been seen for at least a hundred years, with the equivalent of nearly a dozen Nobel laureates. (Math has no such thing as a Nobel Prize, but other honors are considered to confer a similar status within the field.)

"Woodstock" was the closest comparison UCLA math chair Tony Chan could come up with. "It's a once-in-a-century thing."

The official occasion was the hundredth anniversary of a famous lecture given by the great mathematician David Hilbert at an international meeting in Paris in 1900. Hilbert set out the twenty-three most important questions facing mathematicians of his day—thereby setting the agenda, more or less, for the century to come.

The organizers of the meeting at UCLA hoped to do the same.

The range of subjects these rock stars of mathematics addressed was staggering: the mathematics of thinking and subatomic particles, of species and ecosystems, computing and climate change, financial markets and materials, the creation of virtual worlds and the decoding of the human genome.

How does math do it? What is it about this strange hieroglyphics of numbers and symbols that allows it to spread its tentacles through so many disparate fields, holding up the foundation of everything from astronomy to human perception?

"Nobody quite knows," said Chan. "That's why they call its effectiveness unreasonable. People have been trying to explain it [for a long time]. They cannot."

In a strange sense, math can describe everything because it describes nothing. It is an abstraction so extreme that in its pure form, it has nothing to do with the real world. The glory of mathematics, as the late Caltech physicist Richard Feynman put it, "is that *we do not have to say what we are talking about.*" Or as the mathematician Bertrand Russell famously put it: "Mathematics may be defined as the subject in which we never know what we are talking about, or whether what we are saying is true."

Because math isn't stuck in any single context, it is endlessly versatile—like letters of the alphabet, equally adept at writing sonnets or advertisements. Its ability to distill the essence is what makes it universal—like love or hate.

Curiously, however, most math doesn't start out as abstract. Often, it grows out of something very concrete, an attempt to understand some puzzling aspect of the physical world—say, Newton's attempt to understand gravity, which led to calculus.

But then, something strange happens. The abstracted patterns take on a life of their own. "At that point, it becomes mathematics," said Chan.

It can take a while for the math to settle, digest, mature. When it does, the universality of its applications surprises almost everyone: Like the first crude wheel that evolved into cars and trains, ball bearings and salad spinners, Rollerblades and yo-yos, navel rings and *Wheel of Fortune*.

And that blossoming into unexpected forms is precisely what mathematicians think is happening again today and what they intended to celebrate—and build upon—at the conference.

The university also dedicated its new Institute for Pure and Applied Mathematics, one of only three such National Science Foundation–funded centers in the country. "This puts us on the map," said Chan. Previously, people from the East Coast or abroad equated math on the West Coast with UC Berkeley, he said. With the institute, UCLA became a real player. Needless to say, all this was music to the mathematicians' ears. At UCLA, it was time to rock 'n' roll.

Geometry

Mother Nature wears geometry on her sleeve. She spins the stars around in spirals, molds planets into nearly perfect spheres, sends water undulating downstream in sine waves, pulls projectiles into neat parabolas, and holds together the hydrogen and oxygen atoms in water molecules at an angle of precisely 105 degrees.

Geometry even grows on trees: Just look at any flower or leaf.

In fact, nature's thing for geometry has gotten so out of hand, some scientists would say, that she has kissed fields such as physics good-bye for good. Forget forces, particles, fields, gravity, matter, motion—even space and time. Physics has become a chapter in a geometry book.

Today, many physicists believe that everything in the universe—forces, fields, particles, space, time—are merely manifestations of the twisted geometry of eleven dimensions: four make up the extended landscape of space and time; seven curl into origami-like structures too small to imagine, much less perceive.

This set of ideas goes under the name of "string theory." But the name is misleading. These are not strings like pieces of twine. They are not made of anything. They are pure geometry.

String theory seems bizarre, but it is really just the latest in a long line of discoveries that have transformed physics into geometry, matter into shape.

In fact, the idea that geometry rules goes back—like everything else—to the ancient Greek philosophers. Plato's academy in Athens did not admit those who did not know their geometry. "God is a geometer," Plato pronounced.

Aristotle insisted that the planets moved in circles because circles were forms created by the gods. Atoms, to those who believed in them, conferred properties depending mainly on their shape. Fire atoms were jagged, so fire hurt. Water atoms were smooth, so water flowed. Earth atoms were cubical, so earth was solid.

Still, in the Greek universe, at least there was something—that is, matter—to take geometrical form. Geometry didn't rule the universe in its own right.

The real corner turned when Michael Faraday introduced the notion of fields of force in the mid-nineteenth century. Like political spheres of influence, electric and magnetic force fields pervade all space, directing the behavior of everything in their grasp. Iron filings line up around a magnet not because they are being pulled or pushed, but because of a geometrical distortion of space. They follow the curved paths of the magnetic field the way Dorothy followed the yellow brick road to Oz.

Before you knew it, physicists had turned everything into fields: Even so-called particles like electrons are merely kinks in associated fields.

In the early years of this century, Albert Einstein added yet another geometrical dimension to our universe. The dimension of time, he showed, wove together with the three familiar dimensions of space to create four-dimensional space-time. The curving and warping of space-time are behind the force we feel as gravity.

How can geometry be perceived as a force?

Imagine you're a tiny ant living on a paved street covered with potholes. You walk in and out of the potholes unaware of their existence—just as people walk on the round Earth without perceiving its curvature. The geometry of the landscape you live in is simply too huge, relative to little you, to be seen as anything but flat.

Even though you couldn't perceive a pothole, however, you *could* notice that trucks and cars were coming to a grinding halt in the same spot of street. You might conclude that some mysterious kind of "force" was pulling them in—perhaps some exotic rubber-attracting magnet.

In the same way, the "force" we feel as gravity is really just the geometry of space.

"It almost appears that the physics has been absorbed into the geometry," wrote Sir Arthur Eddington, the astronomer who provided many popular explanations of Einstein's theories. "We did not consciously set out to construct a geometrical theory of the world; we were seeking physical reality by approved methods, and this is what happened."

The idea of ascribing forces to the geometry of unseen extra dimensions caught on. In 1919, Polish mathematician Theodor Kaluza proposed that electromagnetism was due to the warping of an unseen fifth dimension. Then Swedish physicist Oskar Klein suggested that this extra dimension could escape detection if it were coiled into tiny, subatomic-scale tubes.

String theory expanded Kaluza and Klein's extra curled-up dimensions into wild new territory. Like Einstein's space-time, the higher-dimensional landscapes of string theory circle into strange, convoluted shapes, forming holes, knots, and handles. And just as the shape of a flute or violin determines the sounds it can make, the geometry of the extra dimensions determines how strings can interact to produce particles. In other words, the geometry determines the ingredients available to make our universe.

Not only do we live in a shapely universe; shape, it seems, is all there is.

"There is still a difference between something and nothing," as Martin Gardner summed it up. "But it is purely geometrical, and there is nothing behind the geometry."

Coincidence

During one week last year, all nine planets clustered together in a small swath of the sky. My new friend's neighbor had a cat with the same name as mine. My stepdaughter in New Jersey and a psychologist friend in L.A. were expecting a baby on the same day.

Were these portentous cosmic occurrences? Or merely curious coincidences?

It's not an idle question. Much of science involves trying to determine which events are connected by cause and which are linked only by the luck of the draw.

"In mathematics, in science, and in life, we constantly face the delicate, tricky task of separating design from happenstance," wrote Ivars Peterson in *The Jungles of Randomness: A Mathematical Safari.*

Is a cluster of cancer cases in a particular town the result of toxic residues, or a normal blip in the distribution of disease? Are those black spots in the photograph evidence that house-sized snowballs are raining on Earth, or random noise in the detector?

Mathematics can help sort out these scenarios. Indeed, a little mathematical know-how may be the only known antidote to a pervasive frailty of the human mind: the perception of causes and connections behind purely chance events.

Human beings are programmed to perceive patterns in chaotic events—the better to make sense of an often confusing visual world. We see forms in cracks on the ceiling, faces on the moon, woolly sheep in clouds. We group clusters of adjacent stars into images of warriors and serpents and dippers in the sky.

"Humans are predisposed to seeing order and meaning in a world that too often presents random choices and chaotic evidence," Peterson wrote.

Indeed, most "amazing coincidences" are created in the mind—like the constellations in the sky—from the human tendency to find plausible links between objects and events.

Consider two strangers sitting in adjoining seats on an airplane. What are the chances that they will have something in common?

Given the vast number of possibilities—from shared favorite authors to hometowns, past lovers and current jobs, schools and acquaintances—it can be shown that in 99 times out of 100 the two passengers will be linked in some way by less than two intermediaries, according to mathematician John Allen Paulos.

Of course, the number of things the two people *don't* share will vastly outweigh the number of things they do. But only the shared links will be remembered as "amazing coincidences."

The same is true, Paulos says, of prophetic dreams. Often, people will dream about, say, an earthquake or a plane crash, only to read in the paper the next day that it actually happened. But given the fact that roughly 250 million people in the United States spend several hours in dreamland each night, "we should expect as much," he says. "In reality, the most astonishingly incredible coincidence imaginable would be the complete absence of coincidence."

Sometimes, coincidences do point to deep truths. Einstein, for example, was bothered by the well-known fact that gravity and inertia balance out exactly in our universe. That is why a bowling ball and a golf ball dropped from a high shelf hit the ground at the same time—because while gravity pulls harder on the bowling ball, inertia endows the bowling ball with a greater resistance to being pulled.

Einstein thought this uncanny equivalence had to be more than a funny coincidence; his realization led him to the idea that both gravity and inertia are aspects of a bigger picture—the curvature of space-time.

In the same way, there is a lot more than coincidence behind the fact that more smokers than nonsmokers get lung cancer. Sometimes coincidence does point to a cause.

How can one tell the difference between coincidences based on cause and those based on chance? Statisticians use all kinds of filters to determine which is which, but even the mathematically averse can use some simple guidelines to keep from getting fooled.

For example, they can remember that large numbers of anything are bound to create coincidences. To wit: It is certain that at least 250 of the 250 million people living in the United States will experience a one-in-a-million coincidence every day—purely by chance.

And while the chance that any one person will win the lottery twice in the same year is tiny, the chance that any of the millions of previous lottery winners will win another lottery in their lifetime is pretty good.

The fact is, events that people normally think of as highly unlikely are often very likely. If someone flips a coin a hundred times, the chances of coming up with a long string of heads or tails are more likely than not. It's not an "amazing coincidence." It's just nature behaving normally.

Or as mathematician Persi Diaconis of Stanford puts it, coincidences are caused by "the world's activity and our labeling of events."

Change

A forty-eight-year-old man who stole over-the-counter drugs was facing life in prison under California's three-strikes law. Critics protested that putting a person away for stealing aspirin was "foolish." But the prosecutor countered that the man was probably a "career criminal... He's going to go out there and continue committing crimes."

People, in other words, don't change.

This is a curious argument—given that about the only thing that doesn't change in this old universe is change itself. Indeed, the understanding that everything changes is one of the most profound insights of modern science.

Only a few centuries ago, people assumed—understandably enough—that the Earth stood still, the stars hung from the sky like sedentary streetlights, and species (including humans) appeared on Earth fully formed as faits accomplis.

Today, we know that the Earth not only spins us around at the dizzying rate of 1,000 miles per hour, it travels 580 million miles a year around the sun.

We know that the seemingly solid ground we walk on drifts about in slow motion, continents colliding like bumper cars. In the process, sea floors ascend to mountaintops and continents sink.

So much for terra firma.

Thanks to the Hubble Space Telescope, we've seen newborn stars just turning on inside celestial clouds—vast interstellar nurseries. We've seen spent stars gasp out their dying breaths in glowing rings of electrically charged gas, exploding like firecrackers, and disappearing into black holes.

Thanks to satellites like the *Cosmic Background Explorer,* we've seen traces of the birth of the universe itself. The 13-billion-year-old ashes still glow at a chilly 3 degrees above absolute zero. We can look back, practically, to where it all began. But the point is, it did begin (the universe, that is). And it's been evolving ever since. Expanding, cooling, clumping into galaxies and clusters.

Even matter has a history. Subatomic particles come into being out of nothing, the pure energy of empty space. Matter changes form, morphing like Jim Carrey from one subatomic species to another. The stars are alchemists, turning hydrogen and helium into carbon and silicon, oxygen and gold.

The very three dimensions of space we live in probably evolved from nine or ten. In the very early universe, space expanded exponentially, inflating the universe like a helium balloon.

For the past 13 billion years or so, it's been expanding at a far more stately pace. But some cosmologists think it's now beginning to pick up speed again, getting ready to stretch the cosmos so thin that we won't be able to see many galaxies much beyond our own. (Not to worry. That's not likely to happen for another 13 billion years or so.)

Time itself, most physicists think, had to begin sometime. The smoothly running flow from past to future that seems so much a part of the given structure of the universe is not, alas, written into the laws of nature. Somehow, the arrow of time evolved.

And what of life? However it started, life as we know it today is unrecognizable from life when it breathed its first on this Earth. For one thing, the earliest life didn't breathe oxygen. Oxygen was a poison exhaled by early organisms. Their deadly fumes extinguished most of the existing life-forms, giving rise to entirely new ones—eventually including us.

The twentieth century has taught us that genes jump around, that brains alter in response to experience, to hormones, to injury, illness, and emotional trauma. Our brains even change as we grow. Recent studies revealed that the brains of adolescents are markedly different from those of adults. No wonder we don't understand them.

But can an individual change? In science, especially, we expect talent and interest to emerge early. The Nobel laureate of today is the toddler who thought deep thoughts and set his home's electric circuitry on fire.

Recently, however, I had dinner with an old friend—a celebrated physics teacher, Paul Hewitt, whose book *Conceptual Physics* has revolutionized the teaching of physics at both the high school and college level. I asked, innocently enough, when he got interested in physics.

You know the punch line: He didn't tune in until age twenty-five. At the time, he was a housepainter with no college education.

Life is rich, strange, and always unpredictable—like the universe itself. But if there's one thing it doesn't stand still for, it's stasis.

Constants

Few things in this changing world are constant. Traditionally, there have been death and taxes, but modern medicine seems to be gaining on the former, and the rich have long found loopholes in the latter.

So it is with the "constants" of physics: Some are a lot more constant than others. In fact, physical constants run the gamut from firmly fixed to highly elusive; several are outright frauds.

At a meeting of the American Physical Society not so long ago, physicists reported new values for the gravitational constant (the strength of the force between two masses), the mass of the electron, and several others too esoteric to easily explain. What actually changed was not nature's value for the constant but the accuracy of the physicists' measurement. In every case but gravity, the new value reflected increasing precision. (Alas, the gravitational force between two masses is so difficult to measure that the uncertainty in its value might have actually increased during the past decade.) But most physicists would agree that whatever the measurements say, the true value of these constants is, well, constant. They've probably always been the same, and they probably always will be.

In contrast are blatantly inconstant constants—for example, the Hubble constant, which measures the rate of the expansion of the universe. You can always find the Hubble constant hiding out at the center of controversies about the size and age of the universe. (If the universe is expanding rapidly, for example, it could have arrived at its present state fairly recently, making it younger than its oldest stars. If it's expanding slowly, then it would have arrived where it is at a leisurely pace—plenty of time for stars to age in peace.)

Probably the only thing all astronomers agree on is that the Hubble constant does not live up to its name. The expansion of the universe has been slowing down ever since the Big Bang— at least until recently, when it apparently started speeding up. In fact, astronomers think the universe expanded so quickly at the very beginning of time that it inflated exponentially—doubling its size thousands of times in less than a cosmic eye blink. Then, somehow, suddenly, this burst of inflation stopped. Erratic behavior for a constant, to say the least.

Perhaps the most notorious constant of all is the so-called cosmological constant. This constant is repulsive in more ways than one: On one hand, it's the energy of empty space that may be pushing the galaxies apart. But it's also, according to University of Chicago physicist Joshua Frieman, "the most maligned constant in the history of physics."

Einstein dreamed it up in order to correct what he perceived as a flaw in his theory of relativity—then later called its invention his biggest blunder.

Now the cosmological constant is once again gaining popularity as a possible fix for several outstanding astronomical puzzles, but it still causes as many troubles as it cures. No one knows where it comes from or even—for sure—if it exists. If it does, it almost assuredly changes over time.

To his credit, Einstein is also responsible for recognizing the most constant constant in physics—the speed of light. "It's hard to imagine how it couldn't be a constant, because it's so embedded in physics," astronomer Kenneth Brecher of Boston University noted at the Physical Society meeting.

The speed of light, for example, is the constant that ties space to time. If it changed, the whole edifice of Einstein's relativity would come tumbling down.

The speed of light is so tied to space and time, for that matter, that its value is nearly impossible to measure. The official measure of space, after all, is the meter—and the meter is defined by the distance light travels in a given span of time. So any measurement of the speed of light is by definition somewhat circular.

What can be measured precisely is light's independence from outside influences. The speed of light is constant whether the person measuring it is running toward the source of light or away from it. And this independence can be measured quite precisely.

In fact, at the meeting, Brecher reported that new measurements using gamma ray bursts from deep space improved its precision by a factor of 100 billion.

In honor of Einstein, Brecher suggested that light speed should be called "Einstein's constant."

Given all the trouble he's had with the cosmological constant, it seems the least physics can do.

Holes

Holes are really something.

We think there's nothing to them, but without holes, coffee wouldn't drip, bread wouldn't rise, baths wouldn't bubble, soda wouldn't fizz. We wouldn't have lace, cameras or parking meters; vases, beer, or bottles to put it in.

A house without holes wouldn't have windows, chimneys, electric outlets, plumbing. Bodies wouldn't have eyes or ears. Clothes wouldn't button. There wouldn't be basketball, bowling, or golf. No volcanoes, no tunnels, no gophers, no caves, no zeros.

A world without holes is inconceivable, and probably impossible. Even the word "impossible" would be impossible without the holes in the letters *P, O,* and *B.*

Some things—like doughnuts—are even defined by their holes. Topologically speaking, a doughnut is the same as a coffee mug, because both have one hole.

Knots are grouped into families defined by the number of holes created as strings loop around each other. In the knotty eleven-dimensional landscapes of string theory, the number of holes is linked to the families of elementary particles.

The person who invented doughnut holes was definitely on to something.

For all their importance, however, "holes are slippery, elusive entities," say authors Roberto Casati and Achille C. Varzi in their book *Holes and Other Superficialities.* On the one hand, holes seem to be real, like rocks, because they have shape, form, and position. We know precisely where a hole resides; it is not some amorphous floating entity, like love.

On the other hand, holes can move. In fact, the motion of holes through the lattice of silicon atoms is the mechanism behind many electronic devices, including computers. The holes are places where electrons are not. They are absences acting as presences; the holes constitute, in effect, positively charged particles every bit as real as the negatively charged electrons that move in the opposite direction. The information is carried, in other words, by holes.

So what exactly is a hole? It's hard to say; even harder, perhaps, to draw. If you're feeling mischievous, the authors of *Holes* suggest, ask someone: Could you please draw a hole for me? Then sit back and enjoy the confusion.

Is a hole a thing? Or the lack of a thing? Is a hole the stuff inside? The boundary? The stuff surrounding? Are the holes in screens made of wire or air?

Clearly, a hole does not entirely depend on what surrounds it, because you can make identical holes in earth, in plastic, in concrete. On the other hand, the hole just as clearly isn't what fills it up; if you stuff your finger into a hole in a dike, it is still a hole—just a filled hole. It is this very fact that "holes are not made of anything," say the authors, that puts holes under "a cloud of philosophical suspicion."

It's even possible to have holes in nothing. Consider a black hole, for example—made entirely of empty space (except for its infinitely small singularity). ?

Or particles of antimatter—which are akin to holes in the vacuum of empty space. Particles can appear out of the vacuum so long as they are accompanied by these "holes"—their ? antimatter partners.

Imagine the vacuum as a flat stretch of ground. If you dig a hole in the ground, you create not only the antiparticle (the hole) but also the particle (a mound of dirt). If you fill up the hole with the dirt, both the mound and the hole disappear. And when particles and antiparticles meet, the "hole" in the vacuum gets filled, and both particles disappear.

There are even holes in perception. People who suffer from strokes sometimes lose track of an entire limb, or even a side of their body. These absences are more than mere erasures, a place where something is missing. Instead, it's as if that whole part of the universe has ceased to exist.

Neurologist Oliver Sacks calls this a "hole in consciousness," just as amnesia is a "hole in memory."

A hole is a place where things leak out and other things come in—like a door. The pupil in your eye lets in the light that enables you to see. But just as important, it blocks out most of the light so that the eye isn't blinded by glare.

A hole does double duty: It is a window and a door; it is something and nothing at the same time. "To talk about valleys is, in a sense, to talk about mountains," say the authors of *Holes*. "But would we eliminate valleys by talking only about mountains?"

In the end, holes are—appropriately enough—holistic. They encompass everything. The hole shebang.

Time

On December 31, my daughter and I were trying to decide whether to bring in the new year at midnight Pacific time, or midnight in New York when the ball dropped at Times Square—or whether, indeed, it made any difference.

And then she asked me: So when did time begin, anyway?

Beginnings, it turns out, pose some of the juiciest questions in all of science.

Consider 2003, for example. When did it begin? When people in New York raised their glasses? People in Tokyo? Siberia? Imagine you could look at Earth from space and watch the new year begin in one time zone after another around the world until the circle was finally complete and it was 2002 everywhere. When every glass everywhere was empty, could you then say the new year had officially begun?

But what about the first year ever? When, exactly, was that?

Given that a year marks one complete orbit of the Earth around the sun, we have to assume that the "year" was born only after the solar system was established.

And when exactly did the clump of dust we now call Earth congeal and settle down into a stable orbit, carving out its first year? Pinning a date on that is problematic at best.

The same is true for the first day. Since the day marks a single rotation of the Earth around its axis, Day One took place when Earth began to spin. But the clumps of dust and rock that made the Earth were spinning well before they joined together. And new spinning chunks (raining from the skies) add their spin to Earth every day.

One thing we know for sure is that the early Earth rotated a great deal faster than it does today, so that Day One, whenever it was, was a whole lot shorter than January 1, 2003. As recently as 900 million years ago, the length of a day on Earth was only about eighteen hours.

Of course, one could argue that it makes no sense to talk about days and years until humans first noticed that patterns of light and darkness fell upon the Earth in regular rhythms, that seasons came and went like tides. When did life become sufficiently conscious to ponder beginnings? Did conscious thought begin with people? If so, when did humanity begin? When did life begin? These questions are so controversial that a dozen researchers might well give you a dozen (or more) different answers.

Curiously, there are also beginnings to things we take for granted as having been around forever. Death, for example.

University of Massachusetts biologist Lynn Margulis has argued that single-celled creatures don't really ever die, because they procreate by dividing in two. They literally live on in their offspring. Only when living things began to reproduce themselves sexually did death of the individual begin. Death, Margulis says, is the price we pay for sex.

Matter, also, has a history. When the universe was formed 13 billion years or so ago, space was filled with energy that cooled into quarks and other elementary particles, eventually producing atoms and the rest.

The energy somehow appeared with the Big Bang—the primordial explosion that set off the universe, life...everything. But when, exactly, did the bang begin?

The question can't really be answered—except to say, perhaps, that it began at the beginning. Since there was no time or space before the universe began, time itself was born along with the universe. The universe began at time zero. No time before is imaginable.

If you find that hard to get a handle on, physicist Stephen Hawking suggests you try to imagine what's "north" of the north pole. There's no there there.

Of course, humans have known since Einstein that the time we measure on our puny clocks and watches is largely an illusion. Time speeds up and slows down depending on how fast things are moving relative to each other—which is why a clock sent on a round-the-world journey by jet returns home running slow compared with a twin clock left sitting still.

Gravity also tugs on time. Time runs slower on the sun than it does on Earth; time at the edge of a black hole stands still. But much is still not known about even the simplest matters of time. For example, why does it flow one way? Why is it that we can travel left or right in space, but only in one direction through time? Space and time, as Einstein showed, are part of a single fabric of space-time. So why should time behave so differently from its partner?

As the poet Octavio Paz put it: "The great lesson of modern science is that...questions about the origin and the end are the most important ones."

And also the most elusive.

PART III

Doing It

Play

Sir Alexander Fleming, the Scottish bacteriologist (1881–1955), had a most peculiar pastime. He liked to paint pictures in petri dishes with a palette of living germs. Being thoroughly familiar with microorganisms—their individual colors, textures, growth rates, and so forth—he was able to produce striking portraits: a mother and child, a ballerina, his house.

Fleming is far better known for his breakthrough discovery of penicillin than for his microorganic art. But he was clearly a man who knew how to play. "I play with microbes," he once said of his work. "It is very pleasant to break the rules."

How sad that so many people seem to have such a hard time being serious about silliness. Even when grown-ups go out to play, their games seem oddly intense and rigid: handball, tennis, running (who skips anymore?), swimming laps. There's a noticeable absence of giggles.

Scientists have always known the value of fooling around. Einstein was famous for his "thought experiments," fantastic flights of fancy that led him to imagine, for example, what it might be like to ride on a light beam, a cerebral magical mystery tour that offered him the insights he needed to produce the special theory of relativity.

"It is striking how many great scientists have incorporated play into their lives and work," wrote Robert S. Root-Bernstein, a physiologist at Michigan State University. "One mental quality that facilitates discovery is a willingness to goof around."

If grown-ups are frightened of play, it's perhaps for good reason. Play is an invitation to break rules. By definition, it's somewhat out of control. There's always the risk of making a fool of yourself.

Only by risking ridicule, however, can one come out from under the covers of conventional wisdom far enough to come up with truly new solutions.

Scientists now complain that this is increasingly impossible. Foundations and federal agencies have become so careful (with a few very notable exceptions) that researchers must submit lengthy, detailed descriptions of the expected outcomes of the experiments or projects they wish to pursue. Ironically, this precludes the discovery of anything unexpected, which, in effect, precludes discovery itself. "Discovering" something you already know is there is like "discovering" the eggs that the bunny hid on Easter morning.

Nature, unfortunately, isn't so cooperative and may hide treasures in the most peculiar places. She may even decide to hide toothbrushes instead of eggs, perhaps in a fifth (or tenth) dimension.

In science, stories of fundamental discoveries made while poking around in the wrong places are legendary: Johannes Kepler (1571–1630) discovered the true elliptical shape of the planetary orbits after devoting a lifetime to trying to prove that they had to be circles. Kepler's method was nothing more than an elaborate game of blocks—trying to fit spherical orbits into cubic (and tetrahedral) holes.

Play is the name we give to the freedom to go out on a limb with the full knowledge that we might fall flat on our faces. In this sense, democracy is a very playful form of government. Making mistakes is built into the system, along with the means for correcting them. Politicians send up trial balloons—the safest way to take a risk, like the child who lobs a fresh remark, then smiles as if to say, "I didn't mean it." Play allows the flexibility required to continually tune responses.

The one place we can all recognize the crucial role of play is in the arts. The writer Annie Dillard took an idea and toyed with it like a cat: for example, the oddity that birds should sing. Perhaps it is a form of bird play, she suggested. Wordplay. Bird word play.

Creativity always comes from such odd juxtapositions. Inventions and discoveries are often based on unexpected combinations and strange connections; sitting in a meeting where silly ideas were tossed about like paper airplanes, occasionally, someone picks one up and makes it really fly.

This can't happen when ideas are proffered, already polished, on silver platters, meticulously packaged in well-researched presentations, too precious to be thrown about. The best ideas rarely come in shiny boxes. They come off the wall.

Off the wall means, simply, coming from somewhere unexpected. Being open to the unexpected is what play is all about.

Failure

Everyone knows about the rewards of success. The stores bulge with (frequently successful) books on how to succeed in business, at school, in love, and even in death.

Why is it that no one talks about the equally impressive rewards of failure?

In the natural and technological world, failure is a critical element in most successful endeavors, be they engineering projects or fundamental discoveries about the nature of the universe.

Consider a stiflingly hot day when the air conditioner blows full blast and two teenagers decide to blow-dry their hair on the same electric circuit. Suddenly, there's an ominous silence. The circuit shuts down. The hair dryers won't start, and the lights won't go on. In other words, failure.

Or is it? Consider what would have happened if the circuit breaker didn't trip under such a power surge. You could have had a fire on your hands, or worse. The blowing of a fuse is the kind of deliberate, purposeful failure that saves us from far worse fates.

"We rely on failure of all kinds being designed into many of the products we use every day," writes professor of engineering Henry Petroski in *American Scientist*. Failure, says the engineer, is often "a desirable end."

Among the ingenious failures engineered into everyday products are the perforations in postage stamps, which allow the gummed paper to give way easily along the tear line. Hoods and bumpers on cars are built to be easily crushed, the better to absorb the impact of a crash. Cracks in sidewalks are deliberately engineered weak points designed to shift as tree roots or earthquakes rattle the concrete; breaking cleanly at the line preserves the integrity of the square.

Petroski also mentions such natural engineering wonders as eggshells. A chick that couldn't peck its way out of its shell would be a dead duck. If the egg didn't break easily, says Petroski, the shell would become "an instrument of extinction" for its inhabitant. Either the egg fails, in other words, or the chicken does.

In a sense, animals that hibernate for the winter and trees that lose their leaves slip into a purposeful "failure mode" that allows them to survive a hostile environment. People sometimes lose their memories after traumatic events—another kind of self-protection. Indeed, psychologists say forgetting is as important as remembering. If we remembered everything, we'd be hopelessly confused. So failing to remember is success.

Of course, chances are human beings wouldn't be around at all if it weren't for the spectacular failure of the dinosaurs— wiped out, scientists think, in the aftermath of a colossal cosmic impact. Much of evolution is driven by genetic "mistakes"—mutations in DNA caused by breakage. Most of these "mistakes" are hazardous to our health, and their progeny soon die out. Others evolve into new, and successful, species.

Indeed, diseases like cancer are created by an overdose of success. Cells that should fail to divide somehow don't get the message, and proliferate out of control.

In science, failure and success often become hopelessly entangled. Many "successful" discoveries or inventions have come about because people made mistakes. Penicillin, like X rays and the glue used in Post-its, was discovered by accident.

To find fertile ground for discovery, scientists often deliberately look for places where the laws of nature appear to break down, where known theory doesn't work, where established models lead to experimentally verified "mistakes." Like fault lines in the earth, these rifts between experiment and theory are often places where a lot is going on.

Einstein's theory of relativity was born in part from the spectacular failure of previous physicists to understand such mysteries as the "luminiferous ether" that supposedly carried light waves through empty space.

These days, physicists fail to understand why 90 percent of the matter in the universe appears to be missing, why the laws that rule nature on the smallest scales seem incompatible with those that rule the largest. Because of these "failures," the best minds in the business have turned their attention to figuring out the discrepancies.

After all, even Columbus discovered the Americas by mistake.

Wrong turns and missteps are not digressions on the road to a predetermined goal. On the contrary, they are often the most successful part of the journey.

Answers

If ½ is the answer, what is the question?

Oddly, finding questions to answers can be a lot more difficult than finding answers to questions.

If the question is, What is the sum of 2 plus 2? the answer is straightforward: 4.

But if the answer is 4, what is the question?

Is it: How much is 2 plus 2?

Or perhaps: How much is 12 divided by 3?

Or even: How many legs on a dog?

Or: The name of a number that rhymes with "door"?

It's one thing to do an experiment that tells you whether or not A or B is the right answer. But given A or B as the answer, how do you find the "experiment" that produced it?

Finding questions to answers gets even harder when dealing with complex systems, such as living organisms. Or as neuroscientist Charles Stevens of the Salk Institute poses the problem: "If a mouse is the answer, what is the question?"

All living organisms are "solutions" to sets of problems posed by the combination of environment, genetics, laws of nature, and chance—to name a few of the factors that come into play. So, given a certain property of organisms, a scientist would like to know: Why is it that way and not some other way? What is the question, in other words, that led to that answer?

In his own research, Stevens is looking for the "question" to ½ because it's an answer that keeps popping up in studies of how the brain processes information. To put it very simply, more brain cells process information than feed information into processing cells. But how many more depends on the size of the animal.

Roughly speaking, the bigger the animal, the bigger the ratio of processing cells to input cells. As you go progressively to species with bigger brain size, the ratio increases. The number that inexplicably keeps popping out of those calculations is ¾.

He'd like to know why.

"We have these laws that work over all animals, from mice to elephants, and it seems we ought to understand what they are."

Looking for questions to answers in the life sciences has been a preoccupation at least since Darwin. The theory of evolution through natural selection is the question to answers ranging from dinosaurs to begonias. (Question: What species produces the most successful offspring under a given set of conditions? In one case, the answer was begonia; in another, T. Rex.)

It's also common in physics. Newton, for example, realized that the orbit of the moon and the fall of the apple were different solutions to the same problem: How do objects behave under the influence of gravity? Codified into equations, the "question" that he discovered was the laws of gravity. The "solutions" range from the parabolic path of water in fountains to the sun's hold on the planets in the solar system.

In the same way, every species of atom is the "solution" to the laws of quantum mechanics. So is every property of matter. Just as a long neck is the giraffe's answer to tall trees, "ice" is the answer to the question: What happens to H_2O if you cool it to below thirty-two degrees Fahrenheit?

By extrapolation, the entire universe is a solution to the equations that describe the laws of nature. If we knew exactly what all those laws were, we could create a universe from scratch.

Or as physicist Leon Lederman frames the challenge in his book, *The God Particle*: "If the universe is the answer, what is the question?"

Complicating the matter is the fact that a single set of questions can have a wide variety of answers. Consider a game of chess, for example, or baseball. In a sense, both are sets of problems to be solved. But even when the rules—and even the players—stay the same, the number of outcomes theoretically possible is infinite.

So why have some games actually been played out while others remain only possibilities? Why did natural selection produce dinosaurs but no Cyclops? Huge flying birds but no flying purple people? Why is the universe created from three dimensions of space and one of time? Why is there gravity at all?

Is there a hidden set of higher laws that determine the result? Or is random chance involved? Or both?

We already know the answer: It's the universe we inhabit.

But the questions are still out there.

Questions

Why? Why? Why?

Some scientists are as pesky as kids, always wanting to know "Why?"

Why this? Why here? Why now? And who ordered that?

Why is the sky blue? Why does time flow forward? Why is there a universe at all?

Even the seemingly simplest questions can set them off. Try answering the "blue sky" question, for example, with the standard explanation: The sky is blue because molecules in the atmosphere scatter the blue wavelengths of sunlight more readily than the redder rays. (The leftover red rays sneak through, reaching us from clear across the Earth at sunset.)

They'll probably just ask: Well, *why* do the molecules scatter blue more readily than red?

If you answer that (the reason has to do with the wavelength of blue light relative to the size of clusters of air molecules), they'll probably ask: Well, why is light the way it is? And why do air molecules cluster?

In recent years, the physicists' whys have gotten more and more outrageous. Twenty years ago, I heard the great MIT physicist Victor Weisskopf lecture a high school class about the "Big Bang"—the beginning of the universe. One student asked him what happened before the Big Bang. Why did it bang?

Weisskopf responded that the "why" of the Big Bang fell outside the realm of science.

Today, the study of why the universe banged into existence is a major question in cosmology. Indeed, one of its leading practitioners, Alan Guth, is the Victor Weisskopf Professor of Physics at MIT. Guth's seminal theory of inflation is really about the tremendous expansion of space that took place before the Big Bang and the creation of the stuff that banged in the first place. "The question of the origin of the matter in the universe," says Guth, "is no longer thought to be beyond the range of science."

Cambridge physicist Stephen Hawking has taken the question even farther: "Why," he asks, "does the universe go to all the bother of existing?"

Cosmologists aren't the only ones asking prying questions of nature. Particle physicists spend billions of dollars to find out why protons weigh what they do, why electrons exist, and why the universe is made of matter.

This insistent digging for deeper truth is not idle theorizing. Often, finding out "why" opens doors to new discoveries.

For example, Newton figured out that the force of gravity is proportional to an object's mass. (In other words, the bigger they come, the harder they fall.) But his theory contained no explanation for this fact. "The gravitational force in Newton's theory might have depended for instance on the size or shape or chemical composition of the body," notes University of Texas physicist Steven Weinberg in *Dreams of a Final Theory*.

Einstein's theory of gravity, on the other hand, explained *why* gravity and mass are related. He showed that gravity is the warping of space-time by matter. Mass creates the warpage that we call gravity. And explaining gravity as the curvature of space-time led to the discovery of a host of exotic phenomena—including black holes.

These days, some physicists believe they've gone even far-ther in unraveling the "why" of gravity. So-called "string the-ory," which explains everything about the universe in terms of strings vibrating in ten-dimensional space, actually requires gravity. A stringy ten-dimensional universe must have gravity. So string theory supplies the answer to the "why" of gravity's very existence.

As you probably guessed, however, other physicists are ask-ing: Why strings? Why ten dimensions?

Even if someone finds a final theory, says Weinberg, it will not explain everything. We'll still have to ask "Why is there any-thing at all? . . . In however attenuated a form, I think the old question—Why?—will still be with us."

That's a good thing for science, because asking "why" is just as important to progress as getting an answer. Maybe even more so. As Einstein put it: "The formulation of a problem is often more essential than its solution."

Weisskopf liked to talk about his old friend and mentor, Niels Bohr, the "father" of quantum physics. Comparing Bohr to a present-day Nobel laureate known for his quick wit, Weiss-kopf remarked that Bohr was a better scientist. The present-day physicist, he said, was brilliant. "He had an answer for everything."

But Bohr was the true genius, Weisskopf said. "He had a question for everything."

Nuisance Value

L ike many big brothers, my son used to think of his kid sister as little more than a nuisance. Then one day, an older boy wised him up: "Be nice to your sister," he said. "One day she'll have money; she'll have friends."

Science, too, has learned how to cash in on the bright side of a nuisance.

Take the lifesaving drug penicillin, for example: It first showed up as an uninvited mold in Alexander Fleming's petri dish.

Or the cosmic background radiation—the leftover glow of the Big Bang. Still raining down on us from 13 billion years ago in space and time, that cool light carries messages about the overall shape, composition, and origin of the universe. But it, too, first made its appearance as a nuisance—an annoying background hiss in a radio astronomer's antenna.

One of the biggest nuisances for accelerator physicists has always been something called "synchrotron radiation." Accelerator physicists are the folks who push tiny subatomic particles around in "racetracks" up to nearly the speed of light. When the particles collide, either with each other or with stationary targets, they can create exotic forms of matter—revealing fundamental properties of nature.

A synchrotron is a kind of accelerator, and synchrotron radiation, in a nutshell, is the light that leaks off the subatomic particles circling around huge rings. The more light that radiates away, the less energy is left in the particle beam to collide with other particles.

In the early days of accelerators, synchrotron radiation was just one of those nuisances—like kid sisters—that physicists had to deal with. For some, it's still an obstacle. In modern machines, in fact, the radiation can be powerful enough to fry sensitive electronics, cause cables to crack, hoses to leak.

But other scientists have long learned how to make a proverbial silk purse out of a sow's ear—using synchrotron radiation as a light source for looking in on the hidden mechanisms of atoms, molecules, and materials.

At first, these "synchrotron light sources" were mere parasites, sucking off the discarded light from particle accelerators; today, they are special-purpose machines, precisely tuned to produce consistent, pure, and highly concentrated packets of light.

The first of this breed to be built was Lawrence Berkeley Laboratory's Advanced Light Source, overlooking San Francisco Bay. As electrons whirl around inside a ring with a diameter two-thirds the length of a football field, they get nudged by powerful bending magnets, as well as "wigglers" and "undulators." And any time an electron is forced to change direction, it sends out radiation, or light. (It's the same wiggling of electrons that produces the light from the sun, as well as the radio waves that bring the sound of your local DJ into your car.)

Once produced, thin slices of the spectrum of light can be diced off and directed to targets at specific experimental areas. The target might be a bit of mud, for example, and scientists might be probing it to see if toxic contaminants are present, and if so, what molecular mechanisms transport pollutants from place to place; or it might be a new material with promising magnetic properties, destined for the next generation of computers; or a molecular complex involved in respiration in bacteria that might help humans breathe easier in the future.

Arriving at targets in precision pulses, these bullets of light can knock an electron out of an atom, or hit an atom with just the right energy to make it ring. They can deconstruct molecules like Tinkertoys, peeling off one layer of electrons at a time to reveal their inner structure.

"You can roam around inside a complex material," as UC Davis physicist Charles Fadley described it. This makes it perfect for unraveling the tangled knots of proteins, for example, or understanding how superconductors work. Fadley's group has recently used the light to produce the beginnings of X-ray holograms that can be rotated and studied from various points of view.

Astronomers even use the ALS to calibrate their instruments, because synchrotron radiation shows up wherever electric charges are pushed around at high speeds—circling Jupiter, for example, or orbiting a black hole, or streaming from an exploding star.

Ironically, perhaps, the ALS was built directly beneath the landmark dome of Ernest O. Lawrence's World War II cyclotron that dominates the Berkeley hillside.

To Lawrence, of course, radiation leaking from his accelerator was nothing but a bother.

Which just goes to show that at least some of the time, life's everyday irritations can have a lot more than mere nuisance value.

Boundaries

In Trivial Pursuit, a question in the science category asks: How many colors are there in the spectrum?

This is the kind of question that can drive a science buff right up the wall. After all, a spectrum is, well, a spectrum— meaning a continuous range. Anyone who has ever looked at a rainbow knows there is no line demarcating where the red ends and the orange begins. Some cultures even clump blue and green into one color. In ancient Greece, there was no word for green. The word *green* actually translated as "wet."

Moreover, the visible spectrum leaks out on each side, bleeding into infrared, microwaves, and radio waves on the low end of the spectrum and melting into ultraviolet, X rays, and gamma rays on the high end.

These deliciously blurred boundaries are typical of most areas of science—which may explain why physicists such as UCLA's Steve Kivelson avoid the science category at all costs when forced into such games.

For example, when astronomers announced that they had discovered the first "brown dwarf" star, others insisted that the "star" was a planet. Still other astronomers call the planet Jupiter an almost-star, while some lump puny Pluto with asteroids.

Even the demarcations between states of matter such as solid, liquid, and gas are not distinct. Glass, for example, is described as solid liquid. The molecules are strewn about in messy disarray, as in liquids. Yet glass feels solid. Given time, however, old heavy glass windows will sag.

Kivelson points out a case he thinks is even more fantastic, in which water vapor under pressure can condense into liquid without going through the usual "phase change," but instead just appears, like a ghost, as water. Like night turning into day, there's no clear line.

"You have no question when you're in night. And no question when you're in day. But [when] do you get there?"

Science thinks nothing of crossing such boundaries. It routinely comes up with strange hybrids such as "organic metals" and artificial DNA. Biophysics—which might once have seemed a contradiction in terms—is now one of the most active fields in science.

Chemists, for their part, have a particular problem with distinctions between "natural" and "unnatural." As Nobel laureate Roald Hoffmann points out, among the human inventions people consider perfectly "natural" are dachshunds and roses—which evolved not by Nature's hand alone but were brought into being by genetic manipulation.

The truth is, most natural phenomena cannot be crammed into clear categories. Where does the sky end and space begin? Where, exactly, is a coastline? Shifting continental plates and erratic riverbeds have made a mockery of property lines and national boundaries alike.

Mathematicians know that even numbers can be indeterminate. The ancient Pythagoreans—who developed an entire religion based on the rationality of numbers—reportedly murdered a member of their clan who threatened to leak to the world their unsettling accidental discovery that some numbers cannot be expressed as ratios or fractions. That is, you can divide the circumference of a circle by its radius but you cannot write the result as a fraction. You cannot pin down pi. Like the Energizer Bunny, it just keeps on going.

It's curious, then, that society often looks to science to draw its lines: to say which fetuses are officially alive, which terminal patients are legally dead—when life and death, like colors, blend into each other. It's the same with most of the other boxes we try to fit things into: who's black or white, who's smart or slow, what traits are masculine or feminine.

These days, many scientists would argue that the most interesting things in nature—from life to the evolution of the universe—exist at the blurred boundary between order and chaos. This is the realm where things are orderly enough to take form, yet not so frozen in place that change can't occur.

It is no accident that some scientists now think life originated in tide pools, at the shifting edge between land and sea.

As the artist Bob Miller of San Francisco likes to point out, there is no graver threat to the process of discovery than that dread disease, "hardening of the categories."

Toy Models

Most physicists and mathematicians are grown-up men and women, so it's surprising to learn how much time they spend playing with toy models.

These models are not exactly the kinds of toy cars or Erector sets or dolls that boys and girls play with. But they do share many characteristics. A model of a car is simpler than a real car, for example. It has fewer moving parts. It doesn't require gas. It's smaller, easier to get a handle on. In the same way, scientists' toy models help them grasp situations too complex to understand directly.

A toy model or theory, according to MIT physicist Alan Guth, is a theory "which is known to be too simple to describe reality, but which is nonetheless useful for theorists to study, because it incorporates some important features of reality."

Cosmologists like Guth obviously need such theories. The kinds of phenomena they study—the origin of the universe in Guth's case—are fraught with unknowns and dependent on dozens of variables. It would be impossible to make progress trying to deal with all of them at once. So scientists selectively trim pieces out of the puzzle until they whittle it down to a manageable size. What they have, then, is a "toy model"—a mathematical description that alters reality to make the problem simpler.

Surprisingly, perhaps, the use of models isn't relegated to complex systems like universes. Discovering the truth about

even everyday things can require creating an artificially simplified view.

Take the well-known fact that all objects fall at the same
rate. In a vacuum, even a feather and a bowling ball dropped
together will land at the same time. But since we don't live in a
vacuum, this simple relationship between mass and acceleration due to gravity is almost impossible to see directly. If you
drop a feather and a bowling ball in Earth's atmosphere, the
ball will drop, but the feather will float on air.

To see the true relationship between mass and acceleration,
Galileo used a model that subtracted the distorting influence
of air.

In the same way, Einstein discovered his first, or "special,"
theory of relativity by getting rid of gravity. As Caltech physicist
Kip Thorne describes it, Einstein "idealized our universe as
one in which there is no gravity at all. Extreme idealizations like
this one are central to progress in physics; one throws away,
conceptually, aspects of the universe that are difficult to deal
with, and only after gaining intellectual control over the re-
maining, easier aspects does one return to the harder ones."

Years later, Einstein eventually figured out how to fold gravity back in, and created his "general" theory of relativity.

Sometimes, scientists create models that eliminate a force
such as gravity. At other times, they'll get rid of a dimension or
two, or simply sweep most of the known particles in nature
under the rug. Any model can work so long as it distills the
problem down to its essence.

The problem is: How do you know that you haven't thrown
the essence out, too? Or made the model so simple that it
doesn't describe anything at all?

That's always a worry. Harvard biologist Richard Lewontin
recently complained about simplified models of human behav-

ior based on data collected from fruit flies. "Always the same story," he said. "You take a simple organism because it is simpler to study, but you factor out everything that's interesting in the process."

Still, models aren't meant to actually solve problems. They're meant to help scientists get a fresh perspective on complex problems. Models, said Stanford mathematician Persi Diaconis, "can be useful in changing our mind-set."

Indeed, scientists will frequently try and abandon many different models before settling on one that seems fruitful to explore. There are many ways to model a universe or a climate system, just as there are many ways to model a person or a car. Cabbage Patch Kid or Barbie, Hot Wheels or Tonka Toys, the variations are endless.

Ultimately, even the best toy models—like supermodels—embody untenable idealizations. They are constructed from ideas in our minds more than from concrete pieces of reality.

As the late physicist Sir Arthur Eddington put it, building such models is not like building a house out of bricks and mortar. "It is like building a constellation out of stars," he wrote.

The stars are scattered more or less randomly in the sky. Humans, contemplating the bright points of light, grouped them into images of dippers and dragons or maidens. But different humans could just as easily group the stars into constellations of cell phones or Hollywood celebrities.

"The things which we might have built but did not," wrote Eddington, "are there just as much as those we did."

Mind and Matter

The campus of the California Institute of Technology sprawls lazily over a small flat stretch of suburban Pasadena, smack against a ridge of high snowcapped mountains. Students in sandals and shorts stroll unhurriedly amid blossoms and palms; a professor pedals by on his bike. Sometimes the silence is so total that you imagine you can hear the scratching of pencils on paper, of chalk on blackboards, the hum of Nobel prize—winning minds.

Three hundred miles north, at the Stanford Linear Accelerator Center (SLAC), the scene, on the surface, is similar. Professors pad about in sandals, and horses graze in the sunshine. Yet beneath the brown-gold hills, giant "atom smashers" accelerate subatomic particles to almost the speed of light, crashing them into each other in collisions that release energies greater than those at the center of the sun, sometimes creating new kinds of matter in the process. Mammoth experimental stations hide in the hillsides, buried in concrete bunkers. The station called Big Mac looked, the day I was there, like a cross between Mr. Wizard's laboratory and the starship *Enterprise*.

Both Caltech and SLAC are at the forefront of efforts to discover the origins of the universe, the innermost mysteries of matter. Yet while Caltech does its work with inspired speculation and subtle math, SLAC probes the subatomic world with elegant technology and sometimes brute force. One comes away from Caltech convinced that the road to the ultimate answers lies in theory; one comes away from SLAC equally convinced that it lies in experiment.

Only in physics is the division between theory and experiment so apparent. SLAC's assistant director of research, William Kirk, told me he thought theorists and experimentalists are entirely different sorts of people—born, not made. "The future experimentalists are the childhood tinkerers," he said. "The would-be theorists are the daydreamers." Indeed, experimentalist Frank Oppenheimer spent his early years taking apart watches and bike brakes, making electric arcs out of house current, and watching lightning storms from the tops of trees. His older brother, Robert, of atom bomb fame, was a theorist and far more cerebral, a better student, a great reader. "He never did learn to split wood," said Frank, "or drive a car."

The work of the theorist is conducted mainly in the brain. Experiments give "hints," wrote Nobel Prize–winning Caltech theorist Richard Feynman, "but also needed is *imagination* to create from these hints the great generalizations—to guess at the wonderful, simple, but very strange patterns beneath them all... This imagining process is so difficult that here is a division of labor in physics." Einstein was a theorist. Theorists tend to be better known than experimentalists, but, said Caltech theorist David Politzer, "that's showbiz. They do the real science, but we get all the attention."

One reason is the numbers. Experimentalists outnumber theorists five to one. As the search for new particles becomes increasingly high powered, it takes more and more people to conduct an experiment—as many as five hundred. Ordinarily, it takes five years to get time on one of America's four foremost machines (like those at SLAC and Fermilab). Theory, on the other hand, is best conducted in the confines of a single mind. Politzer believes in the Volkswagen principle: "Never believe a theoretical paper unless you can fit all the authors into a Volkswagen," he says. "If there are more than that, it means no one person has thought the whole thing through."

Theory requires equipment no more elaborate than a sharp pencil and a prepared mind. Still, it, too, has its drawbacks, said former Caltech president Marvin Goldberger, a theorist: "If the poor theorist walks into his office and picks up his pen and nothing happens, he's stuck. The experimentalist can go and work on a lathe. To be a theorist takes a real single-mindedness. Interruptions are catastrophic. There's no such thing as theoretical physics interruptus."

The power of pure thought in physics is sometimes astounding, even alarming. Quarks and neutrinos and even a property of particles whimsically called charm were discovered theoretically long before they were found in the laboratory. The "machine" with which Paul Dirac discovered antimatter was an equation. Feynman liked to tell the story of the two nineteenth-century astronomers whose calculations of the orbit of Uranus led them to propose the existence of the planet that is now called Neptune. They wrote to two observatories telling them where to look. "'How absurd,' said one of the observatories. 'Some guy sitting with pieces of paper and pencils can tell us where to look to find some new planet.' The other observatory was...well, the administration was different, and they found Neptune."

Needless to say, the experimentalists see things differently. Of course, theory predicts things, they say—but then sooner or later theory predicts *everything*. Wolfgang Panofsky, former director of SLAC, pointed out that while experimental results are often dull reading, "it's the accumulation of knowledge that presents a pressure to find a theory that fits. The discovery of hundreds of particles was the pressure needed to come up with the quark theory." MIT physicist Victor Weisskopf compared progress in particle physics to Columbus's discovery of America: "The experimentalists are those who, when they landed on the other side, jumped from the ship and wrote down what they saw, incredible as it seemed to them. And the theorists are those who stayed back in Madrid and told Columbus he would land in India."

Of course, the division between theory and experiment is not complete. Experimentalists also need imagination, and theorists are not always clumsy with their hands. "I, too, like mathematics," says Frank Oppenheimer. "And Feynman doesn't do experiments but he does play the drums."

Still, the breach is so big that it can take a third species of scientist to bring the experimentalists and theorists together. These so-called phenomenologists are what you might call translators who work between theory and experiment. They have to be intimately acquainted with the intricate mathematics of the theorist, and also at home in the complex cloud chambers and scintillation counters of the experimentalist.

"To some people," said Kirk, "phenomenologist is a dirty word. It implies you are not a deep thinker." As in life outside physics, we tend to assume that those who know more than one thing do not know any one thing well. But in physics, the border between theory and practice provides the perfect breeding ground for bright ideas. "It's cross-fertilization," said Kirk. "The experimentalists like nothing better than to shoot the theories down, so it eggs them on. It's a competition to see who can get there first."

One reason this works so nice is that the specialization in physics is a matter of skills, not content. The goals of both are the same. As long as the lines of communication are open and the progress is complementary, theory and experiment are the right and left sides of the scientific brain. "It's like the oboe player and the violinist," said Oppenheimer. "Each plays a different instrument and requires a unique set of skills. But as long as they listen well to each other, the harmony is fine."

Pomp and Circumstance

Pomp.

And circumstance.

First the pomp: A sparkling Crown Princess Victoria of Sweden gliding down the red-carpeted staircase on the arm of San Francisco mayor Willie Brown in the grand rotunda of City Hall. Champagne toasts by His Excellencies and Her Majesties. Trumpet fanfares and jewels.

Now the circumstance: An inventor discovers a better way to blow things up, decides he doesn't want to go down in history as a merchant of death, so to shine up his posthumous reputation bequeaths his entire colossal estate to work that "benefits mankind."

The man of circumstance, of course, was Alfred Nobel. The occasion for the pomp was the hundredth anniversary of the Nobel Prize, celebrated throughout California with special pride, since the Golden State seems to have a golden touch when it comes to bringing home the heavy gold medals.

It was also an occasion to celebrate the truly singular lesson Nobel left behind—far more important, in the long run, than all the glitter. In short: the power of an individual to change the world.

The prize is nothing if not a personal reflection of Alfred Nobel himself. He directed that prizes be awarded in categories that reflected his peculiar interests: chemistry, physics, and medicine but also literature (he wanted at one time to be an author) and peace (an antidote for his own brand of poison?).

As a scientist, Nobel knew a thing or two about leverage. Since the first Nobel Prize was handed out in 1901, more than seven hundred laureates have been so honored. It's become arguably the most important prize in the history of the civilized world. Sweden and Norway (which gives the peace prize) are small countries that have become pivotal (at least to scientists and peace-brokers) because of Nobel's bequest.

Like most interesting people, Nobel was a complicated man. His early experiments with nitroglycerine killed several people, including his own brother, but he forged on nonetheless. He sold dynamite to armies on both sides of conflicts. "It is fiendish things we are working on," he said. "But they are so interesting as purely technical problems."

At the same time, he cared passionately about peace. In a documentary on the prize, literature laureate Nadine Gordimer tells us Nobel "dreamed of an explosion so powerful that its very existence would make war altogether impossible." So much like the physicists who invented the atomic bomb to stop Hitler. So much like the "war to end all wars."

Did Nobel realize his impossible dream?

War is obviously very much still with us. And the prizes themselves are notoriously skewed. Peace and literature, in particular, tend to get political (nothing new there). Some important fields (like astronomy) aren't represented. Prizes have been given to the right person for the wrong discovery (Einstein did not win for relativity); to the wrong people for the right discovery (let me count the ways); for the wrong discovery (would you believe prefrontal lobotomy?). Far too few women have been honored.

"The only people the Nobel Prize is absolutely good for is

Sweden, my mother, [my university]," said one laureate friend. There's a lot of anguish for the losers, and sometimes a downside even for winners. Suddenly, people are more interested in the laureate than the person behind the prize, my friend told me. "And they ask: 'What have you done lately?' If they ask enough, I begin to wonder myself."

And yet, he thinks the very existence of the prize does much to inspire young people. An old girlfriend recently reminded him that when he was a teenager, he'd told her he wanted to win a Nobel Prize. He did. "Maybe nerdy kids need that," he said. "Especially in a society that finds its heroes on the basketball court."

Many laureates have gone on to become great spokespeople for science, or peace, or literature. That's an enormous service in itself.

I like Nobel's prize for a different reason:

Where else can you find chemistry and physics and literature and peace all celebrated under the same umbrella? Nobel saw that all were of a piece. Wordsmiths, peacemakers, science nerds, stargazers, lab rats—all involved in the same human enterprise: trying to make sense of the universe in which we live. He not only ignored the usual segregation of categories (science or politics or art), he brushed aside national boundaries as well—insisting that the prizes be awarded without regard to nationality.

To top it off, he added an explicitly moral imperative: It is not enough to make discoveries. The prizes go to those who "have conferred the greatest benefit on mankind."

Perhaps some present-day moguls enamored of pomp and lucky of circumstance could take his cue.

After all, as Nobel said: "Contentment is the only real wealth."

Happy Birthday, Walter

A few weeks ago, I received a voice mail message that put a smile on my face a light-year long. The words tumbling out of the machine could barely contain my friend's excitement; he sounded, in fact, a lot like a child who has just got home from school and can't wait to tell you that butterflies come from caterpillars, or that worms can grow their heads back after you cut them off.

Except that my friend was just turning ninety years old and Walter wasn't calling about worms. He was calling about images he'd seen in the *Times* from the newly refurbished Hubble Space Telescope. "They have actually made a picture 120 million *light-years away*!" he said, in a voice still charmingly tinged with his native Vienna. "That's enough to *floor* you. It takes my speech away. It's the *most* important thing that happens today. I'm *very* happy to have heard this."

I, too, had seen the Hubble pictures. But until I heard from Walter, I had forgotten to be properly amazed.

There ought to be a law, I thought. One simply shouldn't be allowed to take such things for granted. Perusing the week's papers, I quickly found a whole new collection of compulsory jaw droppers. Some scientists discovered fossil footprints of footlong "bugs" that had crawled onto earth from the sea 480 *million* years ago; others found a new form of life a mere 400 *billionths* of a meter in diameter, existing improbably in a roiling undersea vent; still others discovered what they think might be a huge reservoir of water on the planet Mars. If there's water, was there life? Is Mom a Martian?

A study showed that genetically speaking, mice and men are barely distinguishable. A telescope produced a detailed image of the universe as it looked 13 *billion* years ago. The cover image of the journal *Science* featured four-dimensional sheets of space-time floating toward each other like tissues in the wind, preparing to collide and create a new universe.

The universe seems to demand that we stay in a state of continual astonishment. Certainly, it is an important quality for scientists.

Take the case of another ninety-year-old who celebrated his birthday last year—Princeton physicist John Archibald Wheeler. He coined both the terms "black hole" and "quantum foam," and now spends his days working on the question of why the universe exists. Talking with a reporter recently, he stopped in the midst of contemplating the connection between reality and thought to ponder a rock on his table. "That rock is about 200 million years old," he exclaimed. "One revolution of our galaxy."

It's harder than you might think to keep such things in proper focus, given the way our minds work. Anything we look at long enough tends to disappear. This can be a difficult effect to notice, because our eyes constantly jitter around, shaking up images imperceptibly, creating a kind of continual visual surprise. If your eyes stayed focused, the visual world would melt.

The same is true of other senses. Whatever drones on long enough goes unheard (if it doesn't first put you to sleep first). We don't feel our shoes until we take them off. The bathtub water stops feeling hot and we stop noticing (if we're lucky) the power lines in the yard.

It's called adaptation, and we're pretty good at it.

It keeps us from getting overwhelmed with information, but it can also make us blind. Before we know it, even science writers forget just how remarkable it is to take a picture of an object in the sky *millions of light-years* away—whatever it is (or was), much changed now. Whatever "now" means millions of light-years from here.

It occurred to me (and I'm certainly not the first) that scientists can be a surprisingly childlike species. Walter, himself, is a retired inventor and engineer who dreamed up some of the fancy optical devices in use at Disney's Epcot Center. But whether the subject is stars or bones or molecules or ten-dimensional space or just great big beautiful machines, scientists are just like kids showing off—sometimes shoving in front of the other kids to get attention.

(Beyond a driver's license, what's the point of growing up, anyway?)

Don Herbert, a.k.a. Mr. Wizard, once told me that he had to use younger and younger children for his program, because once they hit puberty, they become too self-conscious to let their curiosity just hang out. To ask those deceptively innocent questions. The ones Wheeler asks, for example: Is the universe here (or there) when nobody's looking at it? Wheeler thinks not. What's more, he thinks that the act of *looking* is what brings the universe into existence.

So, here's looking at you, Walter.

And Happy Birthday.

An Outrageous Legacy

In physics, as in life, there is a big difference between waves and splashes.

The Edsel made a splash; Elvis Presley made waves. So did Albert Einstein.

So, as was more than evident at a recent sixtieth birthday celebration at Caltech, did one of Einstein's most accomplished successors: physicist Kip Thorne.

Most famously among the general public perhaps, Thorne made a splash with his explorations into the physics of time travel. During a day of public talks in Thorne's honor, Caltech president David Baltimore called him "Caltech's number one strange scientist."

And as Thorne's good friend and colleague Stephen Hawking said during the celebration, understanding time is one of the great challenges physicists currently face. In the spirit of the occasion, Hawking invited Thorne to jump down a black hole—the better to see whether time really comes to an end there, as some physicists speculate.

Of course, as Hawking acknowledged, Thorne would be stretched into spaghetti by the huge gravitational forces inside the hole, and would "not be able to tell us."

Time travel is not a subject for sissies. Colleagues look askance. You can't get funding for time travel outright, as Hawking likes to tell his audiences, so physicists couch it in obscure technical terms, such as "closed timelike curves."

Yet understanding what happens to time in severely warped scenarios where it might break down or reverse sheds light on fundamental unsolved problems. It's hard work because it requires abandoning familiar intuition, hard-won understanding. It means riding into the unknown without training wheels. In recognition of these efforts, Baltimore called his colleague Thorne "the prince of counterintuitive science."

Thorne's research has also made waves in ways that are only beginning to be felt. Thanks largely to the efforts of Thorne, a gravity wave "telescope" named LIGO finally went into operation—the first of its kind. LIGO stands for Laser Interferometer Gravitational-Wave Observatory, and it is, essentially, two pairs of L-shaped four-kilometer-long laser beams strung on opposite sides of the country designed to snare gravity waves from space.

Why would anyone want to catch a gravity wave? Every time a star shudders, it sends out ripples in the fabric of space-time that later lap upon the shores of the Earth. When hugely massive objects like collapsed stars or black holes collide, they crash together like cymbals in the night, making waves. When the universe was born, the Big Bang left its imprint in gravity waves that are still traveling through space today—a whole spectrum of them ranging in size from mere meters to the span of the cosmos.

Encoded in the sounds is a whole new way of looking at the universe. Perhaps, other exotic phenomena not even imagined are calling to us right now. Indeed, when scientists tune into the gravity wave universe later this decade, there's a fair chance that the things they discover will be as startling as those that greeted the first humans to peer at the sky through a telescope.

LIGO, said Hawking, will be Thorne's most lasting legacy. He wasn't counting, of course, the 150-odd "progeny" of

Thorne who gathered together to honor and praise and roast him royally at Caltech, with songs, dancing, and embarrassing memories. Generations upon generations of students and students of students and students of students of students.

Which brings us back to waves and splashes. A splash is a one-shot event. It comes and goes like a shooting star. Its energy is entirely self-contained.

A wave, on the other hand, spreads its influence far and wide—carrying energy and information away from the source, like a rumor passing through a crowd. A wave keeps right on going, ringing out long after whatever started it has gone quiet. It shares its influence with its surroundings.

So it's been, it seems, with Thorne. Speaker after speaker told of his enormous generosity of spirit; his willingness to encourage slow students, sit at a sick colleague's bedside (murmuring encouragement in Russian), deflect credit from himself to others; take a year from his life to find just the right president for Caltech, ten years to write a popular book.

Anybody can make a splash. Witness the *Titanic*.

Thorne, on the other hand, is primed to continue making waves that are sure to last long beyond what is an already impressive (if outrageous) legacy.

Beethoven and Quantum Mechanics

S ome people have such presence that their sudden absence leaves a palpable hole in the universe.

So it is with the death last year of MIT's Victor Weisskopf—not merely a great physicist, but a man colleagues referred to as everything from "the conscience of the physics community" to "the ultimate civilized man."

Viki played chamber music with Einstein ("I've played with better," he said, "but he was very enthusiastic"). He taught quantum mechanics to John Paul II, and loved the idea that a Viennese Jew would advise the pope. (They worked together on disarmament issues.) At Europe's particle accelerator CERN, where he brought about the first truly international physics laboratory, he was known simply as "the Grey Eminence."

To a novice stumbling her way through the confusing maze of modern physics, he was a steady comfort, an eager guide.

I was introduced to Weisskopf through his book, *Knowledge and Wonder,* the only book other than *The Exorcist* I ever stayed up all night to read. Early on, I borrowed his example of how an extra lousy little electron changes the nonreactive gas neon into the highly reactive metal sodium. A fact checker at *Discover* magazine objected that the example I used wasn't technically correct, because sodium also has an extra proton.

When I asked Viki about it, he said: "You always have to lie a little to tell the truth." Years later, when I confessed to feeling guilty because I stole so many of his ideas, he said: "The only sin is if you hear a good idea and you *don't* steal it."

Wrong ideas, he often said, were as important in their own ways as right ones. He told me the story about the impatient German tourist who asked why the Austrians even bothered to publish train schedules, when the trains were never on time. "If we didn't have timetables," the conductor said, "we wouldn't be able to tell how late we are."

When I was frustrated trying to understand gravity as the curvature of space-time, he told me he often felt the same way. He told the story of the peasant who asks the engineer how a steam engine works. The engineer gives a detailed explanation, drawing diagrams, showing where the fuel goes in, how heat is transformed into motion and so forth. When the engineer is done, the peasant says he understands perfectly. Just one more question: "Where is the horse?"

Viki said that's how he still felt about Einstein's theory of gravity. "I understand it perfectly. But I don't know where the horse is."

He said the theory of relativity should be called the "theory of absolutism," and he helped me see that Einstein's genius came from his insight into the unvarying nature of fundamental law. "Relativity says that the laws of nature are absolute and do not depend on the motion of the system," Viki said. "The only things that become relative are old-fashioned concepts like space and time."

He renamed the notorious "uncertainty principle" the "definiteness principle," and passionately defended quantum mechanics against charges that it somehow implied nature was not concrete. On the contrary, quantum mechanics explained why atoms had only a few stable configurations, like musical chords.

All the regularities of nature spring from this astonishing fact—the permanence of genes through generations, the sameness of gold atoms everywhere in the cosmos, the fact that the same flowers bloom again every spring.

He once tried to play the chord for hydrogen on his piano. "It sounds terrible," he told me. "But then, it's not music for our ears."

Above all, he loved the principle of complementarity, developed by his mentor, the Danish physicist Niels Bohr. Complementary truths are seemingly irreconcilable opposites that are both required for deep understanding. "We cannot at the same time experience the artistic content of a Beethoven sonata and also worry about the neurophysiological processes in our brains," he said. "But we can shift from one to another." He told a story about a conversation that took place during a walk on the beach between Nobel laureates Felix Bloch and Werner Heisenberg. Bloch was expounding some new theories about the mathematical structure of space. Heisenberg responded: "Space is blue, and birds fly in it."

Whenever one way of thinking dominated, Viki warned, it spelled disaster for human society—whether it was the dominance of religion during the Middle Ages, or the dominance of technology today.

He liked to say he lived for Beethoven and quantum mechanics. "What's beautiful in science is the same thing that's beautiful in Beethoven," he once told me. "It connects things that were always in you that were never put together before.... There's a scientific truth and an emotional truth. If you are completely immersed in one, many important aspects of life are missed."

Luckily for me and everyone who knew him, Viki didn't miss a thing.

Alan and Lucretius

A lan Alda is a lot like Lucretius.

To be fair, Alda, who played the late physicist Richard Feynman in the play *QED*, is a lot younger than the ancient Roman sage.

Lucretius lived more than two thousand years ago and wrote a famous poem that proffers everything from advice for the lovesick to a theory of the universe. His work, *On the Nature of Things*, translated mostly obscure Greek thought for a popular Latin audience. Today, we might call him a popularizer of science.

Lucretius composed his work primarily because he was concerned about the misery that ignorance and superstition bring upon the world. In the opening of his poem, he tells the story of a young girl "struck dumb with terror" and "led trembling to the altar" as she prepares to "fall a sinless victim to a sinful rite, slaughtered to her greater grief by a father's hand, so that a fleet might sail under happy auspices."

Such, concludes Lucretius, "are the heights of wickedness to which men have been driven by superstition."

Alda isn't necessarily worried about ritual human sacrifice, but he is concerned that many people lack the kind of knowledge of the natural world that nurtures rational thinking.

That's one reason he was attracted to the character of the late Caltech physicist Richard Feynman, whom he portrayed to sold-out crowds in both L.A. and New York. Feynman didn't think you could fool Mother Nature, and he had no patience for people who tried. In fact, he once defined science as "a long history of learning how not to fool ourselves."

"I was completely taken by the Feynman character, and I was especially taken by his courage," Alda said. "He doesn't want to fool you, and he doesn't let himself be fooled."

Alda came to his passion for science the hard way. From the age of six, he was doing experiments, mixing toothpaste and baking powder to see what would happen. "Thank god I was too short to reach things that might really explode!"

At the age of ten, he dreamed up a malted milk machine, a six-armed can opener, and a lazy Susan designed for the inside of a refrigerator.

Later, however, he went through a period when he "thoroughly avoided" science, he said—in part because "the idea was in the air that if you're interested in art, you can't be interested in science."

For a while during his twenties, he even got seriously into the study of clairvoyance and channeling. When the claims didn't hold up, however, he went back to science, this time in earnest, reading *Scientific American* regularly, as well as whatever books he could find.

And then, he said, a funny thing happened: "I noticed my thinking started to change. I got a different idea of how you evaluate what's true."

He's been stuck on science ever since, creating ten hour-long programs a year for *Scientific American Frontiers*. And after years of effort, he finally won the opportunity to channel the spirit of Feynman on stage.

Alda was attracted to Feynman's insatiable curiosity and bottomless lust for life. "He was so curious about everything;

he really believed that everything was interesting if you looked at it closely."

Alda was not above throwing around his silverware and sugar packets during lunch to experiment with gravity, as he did at a Westwood restaurant with a certain science writer who prefers to remain anonymous.

"But we also have this very serious task of promoting understanding," he said. "How do you figure out what to believe?"

Many people in our society lack the tools they need to tell the difference between established knowledge and the gossip they hear in their local health food store. Discerning the truth doesn't require that we all become scientists, Alda said. But it does require that we base our beliefs about the world on investigation rather than faith. To drive his car, Alda said, he doesn't need to know every detail of how it works. "But I like to know where the battery is."

And learning how not to be fooled, he said, takes more than making scientific knowledge accessible. "The tools have to be made accessible," he said. "Otherwise, you have to take things on faith."

Specifically, he wished he'd learned more math. There's still time, the anonymous writer pointed out. Alda was excited because a friend at Cornell had promised to teach him some "real" math. "Maybe I could learn to do an equation!" he said.

Lucretius would have approved.

"This dread and darkness of the mind cannot be dispelled by the sunbeams," Alda's predecessor wrote, "but only by an understanding of the . . . inner workings of nature."

The Sun Painter

The physicist Frank Oppenheimer used to say that artists and scientists are the official "noticers" of society—people whose business it is to notice things that other people never learned to see or have learned to ignore.

I've never known anyone with quite the knack for noticing as San Francisco artist Bob Miller. Since I've known him, countless things I used to think quite ordinary have been animated by his imagination. Once he asked me: How would you suspend 500,000 pounds of water in the air with no visible means of support?

Answer: Build a cloud.

Needless to say, clouds have never looked quite the same to me since.

Neither has dawn. Before I met Bob, I never thought about the continual wave of people rising from their beds that sweeps around the globe as the shadow of night lifts time zone by time zone, much like the wave that runs through a crowd at a ballgame, except that the tide of waking people is a wave that never ends.

I think of this wave as a kinetic sculpture that Bob placed permanently inside my head.

(The U.S. Patent and Trademark Office didn't get this concept at all. One of Bob's sculptures is an optical illusion in which a concave object appears to pop out and follow you as you walk by. The sculpture is made up of an inside corner of a box that appears to turn into a cube, and so he calls it his "Far Out Corner." Oppenheimer encouraged him to patent this optical illusion/sculpture many years ago. They were both amused when the request was turned down. The reason? The bureaucrats maintained that he couldn't patent an effect that existed only in someone's mind. As if there's a piece of art or science that doesn't. The patent office later relented.)

Bob's signature piece is probably his "Sun Painting," which begins with a beam of plain white light from the sun, refracted through a rack of prisms, then sliced with thin mirrors into a palette of pure colors—Bob's "paints."

The late poet Muriel Rukeyser was inspired by this work to write a poem for Bob, which she called "The Sun Painter." She describes standing in the work as being embraced by "a lashing of color. Not color, strands of light. Not light but pure deep color beyond color...You have invited us all. Allow the sunlight, dance your dance."

Before I met Bob, I labored, like most people, under the misconception that the green of the grass, the red of the car outside my window, the yellow of the neighbor's cat, came from the objects themselves. But no, they come from the sun: The purple of that Laker flag is a gift from a star 93 million miles away.

In my mind's eye, Bob "painted" white light in brilliant colors as surely as Lewis Carroll's playing cards painted the Queen's white roses red.

He's done the same thing with shadows, which I used to see as black and white. Now I see a shadow as a "negative" image that contains as much information as a "positive" image. Bob makes a spellbinding case for this in a piece of performance art called "Bob Miller's Light Walk."

Here's a taste. A pinhole camera (in essence, a box with a

pinhole poked in one end) can make an image of any scene, moving or static. But suppose instead of a hole, you use its complement: not the hole in the doughnut, but the doughnut "hole" you eat; not a pinhole, but a pin "speck." The shadow cast by the speck is a complementary image.

Astonishingly, if you dangle a pin speck, you don't just block light, you can create a perfect negative image of a scene.

During a solar eclipse viewing in the Black Sea, after Bob wowed astronomers with his light tricks, he did the same for artists that night sitting in a bar. Using a small piece of cheese suspended on a hair as a pin speck, he cast colored shadows of some brightly colored lights. The shadow image appeared not only upside down and reversed, as you might expect, but also in complementary colors: The red turned to turquoise, the blue to yellow.

Creating a pin speck image can be tricky, especially in a bar. It requires just the right lighting and geometry to make its magic work. But everyone can see the effect during even a partial solar eclipse: The tips of your fingers serve as pin specks that cast clear complementary images of the sun—dark crescents decorating your hands like claws.

To Bob, there's no such thing as uninteresting, diffuse white light. It's awash with images, bursting with brilliant colors; they need only to be separated out, like voices in a crowd. Whether we do it with holes or specks doesn't matter. (Making lifelike portraits out of holes and specks is one of Bob's specialties.)

Bob often says that the worst disease afflicting humankind is "hardening of the categories"—our futile attempts to cram things into boxes and keep them there. Boxes like "art" and "science," for example.

These categories become blinders that prevent us from

noticing the weight of clouds, the colors of white, the light in shadows, the opportunities in obstacles.

Bob's "artist's statement" sums it up: "Blobs, spots, specks, smudges, cracks, defects, mistakes, accidents, exceptions and irregularities are the windows to other worlds."

Invention and Discovery

It's a question as old as Plato: Do scientists discover laws of nature, or create them?

Did Einstein discover relativity, or did he think it up?

Do mathematicians invent theorems and proofs, or are these truths out there waiting to be discovered? Do chemists find new molecules, or forge them?

Sometimes, the answer is obvious. Astronomers didn't create the stars, and physicists didn't invent gravity. Something out there shines; objects fall.

Yet science's long-running creation controversy has deep and tangled roots.

According to Plato, numbers and geometric forms exist as independent entities, "out there," regardless of whether anybody ever discovers them. Like tiny organisms seen only through microscopes, or distant extra-solar planets seen only with the help of telescopes, mathematical truths are simply there to be found.

Today, few mathematicians are strict Platonists. Certainly, some mathematical truths are simply "out there," according to Ronald Graham, a mathematician at UC San Diego. "Perfect numbers have a certain form."

But proving that something is true, he says, "is a creative and elegant thing."

Even numbers that we take for granted today were clearly "created," not discovered. Take zero, for example. The idea of a "number" signifying nothing didn't exist until less than a thousand years ago. It didn't exist until the Indians, the Mayans (and probably others), invented it.

In the same way, calculus was dreamed up in the minds of Newton and Leibniz—along with the bizarre idea of a number that could be infinitely small.

Indeed, mathematician Reuben Hersh argues in his book *What Is Mathematics, Really?* that mathematicians invent every bit as much as they discover—loath as they often are to admit this.

"The majority of those who ponder about mathematics no longer believe in Platonism," writes Marcia Ascher in her classic text, *Ethnomathematics.* "To us, squares or right triangles or prime numbers are categories that we, in Western culture, have created [not discovered]."

Mathematicians have been tossing this question around for millennia. What about other sciences?

Certainly, proponents of string theory—the idea that the universe and everything in it results from the vibrations of tiny strings in ten-dimensional space—are generally adamant in their claim that the theory is being discovered, not created.

"Nobody in this field is clever enough to have invented something like this," says Harvard physicist Andrew Strominger. "It's clearly something that we discovered. It's beyond our imagination to invent such a beautiful and powerful mathematical structure."

But others hold different views.

"Is the universe already there—a single entity . . . waiting out there for us to discover?" asks Colgate University astronomer and anthropologist Anthony Aveni. "Or are there infinite ways to piece the cosmos together?"

His own answer is obvious in the title of his book: *Conversing with the Planets: How Science and Myth Invented the Cosmos.*

Clearly, the process of discovery evolves from a complex equation. Gravity existed before Newton came along, but Newton created the laws that describe how gravity works. Then Einstein invented (or discovered?) a new set of laws (relativity) that describes how gravity comes to be.

Perhaps scientists would view creation differently, argues chemistry Nobel laureate Roald Hoffmann, if science had a different—decidedly less sexist—history.

"The male metaphors of peeking, unveiling, penetrating are characteristic of nineteenth-century science," he says. "They fit the idea of discovery."

But discovery, he says, is only a small part of what scientists do—even though the cliché, "'uncovering the secrets of nature,' has set, like good cement, in our minds."

Hoffmann is particularly upset that chemists, of all people, have accepted this metaphor. Some molecules, he allows, are already there, waiting to be discovered, like the Americas.

"But so many more molecules of chemistry are made by us.... We're awfully prolific." Millions of previously unknown molecules—molecules that never existed before on Earth—have been created like works of art in the laboratories of chemists.

There's a lesson here, says Hoffmann: "Had more philosophers of science been trained in chemistry, I'm sure we would have a very different paradigm of science before us."

They might have to invent discovery all over again.

Red

There's been a certain amount of snickering in astronomy circles over the "color of the universe" brouhaha. In case you missed it, astronomers from Johns Hopkins announced that the universe was turquoise. A month later, they found a bug in their code, and now they say it's closer to beige ("cosmic latte" is the current favored name).

"Why just the color?" one astronomer cracked. "Why not its texture? How does it feel?"

Is it sticky or smooth? Sad or gay? If it's on the green side, does that means it's jealous of some other universe? If it's red, is it blushing? If it's beige, is it bored?

To be sure, there's a serious reason to study the color of the galaxies: spectral lines in the light from stars reveal the mix of ingredients that make them up.

No, what raised eyebrows was the fact that the astronomers only figured in the color of the universe as it would seem to human eyes—an awfully anthropocentric way to look at things: *me, me, me.* The universe would look quite different to a snake or a cat or a bee or a bat.

Focusing on visible light, in fact, means throwing away 99 percent of the electromagnetic spectrum. Astronomy depends on a wide array of telescopes that look at *all* of the light, from gamma rays to microwaves. And if you took *all* the wavelengths from all the stars and galaxies and other astrophysical objects into account, the universe would probably radiate in infrared. Even that light would be swamped by the microwave glow left over from the Big Bang.

Of course, most of the matter in the universe doesn't give off light at all. So the true color of the universe has to take into account the 90 percent of matter that is "dark," or more accurately, transparent. That would make the universe the color of clear glass.

And cosmologists are becoming convinced that most of the universe isn't even made of matter, but instead of some mysterious "dark energy" that behaves like a kind of anti-gravity, pushing galaxies apart. Color it repulsive.

In truth, the Johns Hopkins astronomers hit on something far more subtle and interesting than the color question, which they admittedly did as "a bit of fun," as one of them put it, attached to a larger paper. "One moral of this story is we should have paid more attention to the 'color science' aspect," he said.

In fact, how we see color is a complex and controversial question: "The history of the investigation of colour vision is remarkable for its acrimony," writes Richard Gregory in his classic *Eye and Brain*. "The problems have aroused more passion than passion itself."

Color vision is completely counterintuitive. In the first place, the human visual system evolved to do more with less (rather unlike our economic system). It's been known for nearly two hundred years that we see the entire panorama of possible color with just three kinds of receptors—those for red, green, and violet.

Colors like yellow are mixtures—in this case of red and green. The sensation of "yellow" is something like H_2O, a mixture that is nonetheless pure—and with properties entirely different from the ingredients that make it up. So if you shine a red light and a green light on a white wall, the place where they overlap appears yellow. Add violet to this mix, and the colors immediately bleach into white. No matter how many times I've done this trick, it still astonishes me. (Definitely try this at home.)

Still, the wavelength of light is only part of the story. Seminal experiments by Edwin Land showed that it takes only two narrow bands of light to reproduce nearly the entire visible spectrum in the brain. The color we see depends on everything from general surroundings to expectation, from edges to illumination, physiology and psychology alike.

As Gregory concludes: "Any simple account of colour vision is doomed to failure."

The trickiest color of all is white, and it's the one that got the Hopkins astronomers into trouble. There's no exact formula for "white"; the brain constantly recalibrates white depending on context. Normally, what we see as "white" is just the whitest thing around. If you take a white piece of paper into a room illuminated only in red, for example, the paper still looks white even though only red wavelengths can be reflected from the paper to your eyes. The brain needs to keep colors constant under different lighting conditions; otherwise, the world would be a kaleidoscopic blur.

Unknown to the Hopkins astronomers, the freeware program they used to calculate the color of the universe set a "white point" for a reddish environment. This made the color of the galaxies appear, on average, turquoise.

As the universe ages, there will be fewer young, hot, blue stars, and more older, cooler, reddish ones, pushing the universe farther into the red. Eventually, the stars themselves will collapse into black holes.

Only then will the universe finally find itself in the black.

Objectivity

There it was in black and white: On July 4, 1984, the *New York Times* proclaimed that physicists had finally found the sixth quark—the last member of that mysterious family of elementary particles that buzz around inside the atomic nucleus. The announcement conjured up images of intent men and women in white peering into their apparatus and saying, "Aha! There it is! The sixth quark! What a beauty!"

By chance, a physicist friend from California was visiting that day—a young MacArthur Foundation "genius" who's credited with discovering the nature of the "glue" that binds quarks together. He didn't seem impressed by the announcement. In fact, he was somewhat amused. You see, he explained, the machine that had "found" the quark had been shut down for more than six months. Most of the data from the experiment had been analyzed at least three months earlier. The results were well-known. "What the announcement means," he said, "is that they finally agreed on what they saw."

This story has much to say about the nature of scientific knowledge. It is not, as people are so often told, a collection of objective facts and unbiased observations that sprout in hermetically sealed environments, unsullied by human minds and hands. "On closer analysis," writes science historian Paul Feyerabend, "we even find that science knows no 'bare facts' at all, but that all the 'facts' that enter our knowledge are already viewed in a certain way." Facts come clothed in history, colored by context. Science is less a statement of truth than a running argument. As it turns out, the scientific method isn't so scientific at all.

Heresy? Not at all. Most scientists would probably agree with physicist Robert March of the University of Wisconsin that rules of objectivity only apply to the way facts and ideas are tested. Discovery is quite another matter. "Time and time again," March has written, "a remarkable pattern of discovery has repeated itself: a lucky guess based on shaky arguments and absurd *ad hoc* assumptions gives a formula that turns out to be right, though at first no one can see why on earth it should be."

Ideas like Einstein's relativity, Bohr's atom, and Copernicus's sun-centered solar system went against the grain of most then-accepted theories—not to mention most common sense. "Copernicanism and other essential ingredients of modern science survived only because reason was frequently overruled in the past," Feyerabend has written.

The truth is that science couldn't be completely objective even if it wanted to be. Take Galileo, turning his gaze toward Jupiter for the first time through his telescope on January 7, 1610. Lo and behold, what should he see but "three little stars"—moons of Jupiter! Galileo's colleagues didn't pat him on the back for his remarkable discovery. Much more sensibly, they thought he was seeing an optical illusion. After all, the lenses in a telescope *distort* images (as does the lens in your eye). In Galileo's day, they produced rainbow rings around everything one observed. So wasn't it more likely that the visions Galileo saw—craters on the moon, moons around Jupiter— were also such distortions, rather than reflections of the clear objective truth?

Now compare Galileo and his simple apparatus with modern-day astronomers who stargaze through arrays of telescopes monitored by computers, and you see why you have a problem.

But these are mere technicalities. The real problem is that even the most objective researchers have to focus before they can see, which means they have to decide where to look. This limits their options. Even theorists must focus their thoughts. As Einstein said, "If the researcher went about his work without any preconceived opinion, how should he be able at all to select out those facts from the immense abundance of the most complex experience, and just those which are simple enough to permit lawful connections to be evident?"

Preconceived notions are the enablers of scientific discovery. In fact, for MIT science historian Thomas Kuhn—who has probably written more about this subject than anyone else—normal scientific research is "a strenuous and devoted attempt to force nature into the conceptual boxes supplied by professional education." This isn't a bad thing. These conceptual boxes are like telescopes and microscopes that let researchers zero in closely and therefore "penetrate existing knowledge to the core." Preselection is an essential step in the process.

At the same time, preconceptions certainly change the way scientists (and other people) see things. Kuhn makes this point by describing a psychological experiment that deserves, he argues, "to be far better known outside the trade."

The experiment is simple enough: Subjects are asked to identify playing cards. Most of the cards are normal, but some are anomalous—— a red six of spades, for example, or a black four of hearts. Subjects usually have no trouble whatsoever identifying the normal cards correctly, but they also identify the anomalous cards as if they were normal. "Without apparent hesitation or puzzlement," writes Kuhn, they will identify a black four of hearts, for example, as the four of spades. Only after many exposures do they realize that something unusual is going on.

Kuhn compares this to the way scientists perceive (or don't perceive) the unexpected in nature. "In science, as in the playing card experiment, novelty emerges only with difficulty, manifested by resistance, against a background provided by expectation. Initially, only the anticipated and the usual are experienced even under circumstances where anomaly is later to be observed."

Like the rest of us, scientists tend to see what they expect to see. Darwin once spent a whole day in a river valley and saw "nothing but water and plain rock." Eleven years later he walked in the same valley, this time looking for evidence of glaciers. "I assure you," he wrote a friend, "an extinct volcano could hardly leave more evident traces of its activity and vast powers.... The valley about here must once have been covered by at least eight hundred or a thousand feet in thickness of solid ice!" Once Darwin knew what to look for, it was easy enough to find.

Scientific objectivity is inevitably blurred by the biases built into human perception. So perhaps it's the human element that ought to be eliminated. Perhaps we should let machines do our science for us. Wouldn't that make research more objective?

In some simple senses, yes. (Computers can certainly keep better track of subatomic encounters than people can.) But when it comes to interpretation, a computer is biased by the very nature of its "thought processes" (not to mention the thought processes of programmers). Computer thinking is clear, linear, logical. The real world (even the real mathematical world) is tangled, paradoxical, complicated. By the time programmers translate ideas into algorithms that computers can assimilate, they've already made scores of assumptions.

And even when computers do manage to spew out "completely objective" facts, they're often of little interest to scientists, who tend to find naked facts fairly useless. Lewis Thomas, for one, defined a scientific fact as such only "when the data, taken together, mean something." He liked to quote Albert Szent-Györgyi's remark that "discovery consists of seeing what everybody has seen and thinking what nobody has thought."

How is it, then, that for most people, "scientific" has come to be nearly synonymous with "objective"? Historians such as Kuhn and Feyerabend blame it on how science is taught and presented to the public. "A little brainwashing," wrote the latter, "will go a longer way in making the history of science duller, simpler, more uniform, more 'objective,' and more easily accessible to strict and unchangeable rules."

This brainwashing, Feyerabend concluded, does a terrible disservice to science. Acknowledging the essential subjectivity of science makes it stronger because it keeps the door to reinterpretation open. Accepting a single, objective, truth, on the other hand, means that someday the door will be closed. "Unanimity of opinion may be fitting for a church, for the frightened or greedy victims of some (ancient or modern) myth, or for the weak and willing followers of some tyrant," he wrote. "Variety of opinion is necessary for objective knowledge."

So is science essentially a sham? Is one person's opinion as good as another's? Is there no such thing as scientific truth?

On the contrary, science earns its reputation for objectivity by treating the perils of subjectivity with the greatest respect. That's why science is so sticky about testing (and retesting) results. The true measure of scientific objectivity is consensus, as my MacArthur friend pointed out.

In fact, farsighted scientists always try to keep an eye out for the distortions that are inevitably created by the lens in the mind's eye. A few months ago, I was working with another MacArthur fellow who was exploring theories of strings that vibrate in ten-dimensional space. He constantly worried, he said, that he was like the man who looked under the lamppost for his keys because that was the only place he could see. True discoveries are almost always made while stumbling about in the dark.

And P.S.: The reported discovery of the sixth or "top" quark turned out to be premature after all. The quark's existence was finally established by subsequent searches in February of 1995.

Wherever You Go, There You Are

Isn't it strange that scientists went all the way to Mars—only to discover that the rocks there are named for familiar cartoon characters on Earth? Scooby-Doo, Casper, Calvin and Hobbes.

Alas, it's an old story in science. The late physicist Arthur Eddington told of the astronomer who gave a lecture on stars. Afterward, a student complained that while he could understand how astronomers had discovered the temperatures, masses, and distances to the stars, he couldn't understand how the scientists found out their names.

We chuckle, because we know that people give the stars their names; we know that people—even scientists—invent aspects of the outside world based on the images already in their heads. When Mars Pathfinder scientists reported that the Sojourner rover "did a wheelie, it was so excited," we know that the robot wasn't excited; the scientist was. Yet it's easy to confuse the world that is with the world we make up.

For example, as the year 2000 approached, many people were obsessed with the idea of the approaching millennium. The year 2000 (or 2001, more correctly) seemed to signify something profound. But there's nothing special about a thousand. It's an arbitrary point in time with no meaning at all in the natural world.

In fact, the whole idea of time is just such an artificial concept. When you cross the international date line, your plane doesn't go bump in the night. Time only tells us the position of Earth relative to the sun, or the number of sand grains that have run through the egg timer, or the number of vibrations of a quartz crystal in a watch.

People made up the millennium just as they made up seconds and minutes—and, for that matter, numbers. The number zero, in fact, is a fairly recent human invention. It makes much of mathematics easier, but ten apples plus ten apples is twenty apples whether we have zero in our number system or not. We could say the same thing in Roman numerals.

Tools like numbers are useful fictions. They play an invaluable role in revealing connections and discovering aspects of the physical world. They bring a semblance of order to the universe, making it simpler, clearer, easier to grasp.

Likewise, Scooby-Doo is easier to remember than rock #xpz15r. But the name tells us more about ourselves than about the rock.

Consider the constellations. The stars that we stitch together into outlines of dippers or twins or warriors may be far removed from each other in space. We connect the stars for our own convenience.

Tellingly, the constellations in the Northern Hemisphere resemble gods and beasts and royals who mattered to the Greeks who named them, while southern constellations reflect the more modern interests of the first global navigators—instruments and geometric figures.

"The division of the stars into constellations tells us very little about the stars," wrote the late physicist James Jeans, "but a great deal about the minds of the earliest civilizations."

Some of the useful fictions scientists invent to help them get a handle on the familiar world later turn out to be real. Molecules, quarks, and electric charges all started out as metaphors for the unfathomable. Only years later did technology allow scientists to see that they actually existed.

Albert Einstein invented just such an arbitrary factor to make his equations mesh more smoothly with reality. He called it the "cosmological constant," and he later said it was the biggest blunder of his life. Lately, however, the cosmological constant appears to be making a comeback. Cosmologists think it might be a real kind of antigravitational force that accounts for the so-called dark energy that dominates the universe.

It's a fine line, sometimes, between discovering something and making it up. Scientists continually have to pinch themselves to see whether they're really looking into a mirror when they think they're looking through a window at the outside world.

Nobody said it better than Eddington: "When science has progressed the farthest, the mind has but regained from nature what the mind has put into nature. We have found a strange footprint on the shores of the unknown. We have devised profound theories, one after another, to account for its origin. At last, we have succeeded in reconstructing the creature that made the footprint, and lo! it is our own."

Metaphor

An old friend wrote me recently from his new post at MIT, describing one of his colleagues as a typical "dot edu," and another as an uptight "dot com." As practically everyone knows by now, the dot followed by the three letters designates the origin of an Internet address: edu for educational institutions, com for commercial enterprises, org for nonprofits, and gov for you-know-who.

Inevitably, the jargon of science invades the popular lexicon. We've grown accustomed to calling suggestions "input," and dismissing ideas that "don't compute." These days, if a physicist wants to send you something, he's likely to ask you to give him your "coordinates."

We tend to forget how much of our everyday speech descends from science. Timothy Leary's call to "turn on and tune in" wouldn't have made much sense before the invention of the radio. Nobody talked about taking "quantum leaps" (much less named a TV show after them) before the inner world of the atom revealed its quantum mechanical secrets. And no one was described as "irrational" before the discovery of irrational numbers.

Metaphors also flow the other way, from everyday life to science. Indeed, metaphors are scaffolding that holds science up while new theories are under construction. Fields of magnetism are not like fields of daisies, but they help us visualize how influences can spread through space. Electrons do not flow through wires like water through faucets, but thinking of currents this way helps us to imagine how electricity might work.

Eventually, the metaphor breaks down, the scaffolding cracks under the weight of new evidence. These cracks are good places to look for new physics.

Take the atom. Before its inner structure was understood, physicists described it as a miniature solar system—electrons orbiting like planets around a central nucleus. But they soon realized that atoms didn't behave like planetary systems in several important ways. For example, every carbon atom (or helium or iron atom) in the universe is exactly alike.

How do atoms remain forever the same, like some minuscule Dorian Gray, without the benefit of plastic surgery?

The trick is, they absorb or emit energy only in precisely measured parcels (the famous quantum leaps). The subatomic world is not like a solar system at all, but rather like stadium seating, with electrons confined to moving in specific rows. If the atom were jostled or torn apart, the electrons could still only "sit" in given rows, meaning the atom could only come back together in one of a few preset configurations.

Of course, the stadium metaphor is ultimately inadequate as well. It may help us visualize why atoms absorb energy in lumps, but it doesn't explain why they also change their "spin" in lumps. Indeed, it's hard to imagine how particles such as electrons "spin" at all, given that they are pointlike, with no dimension. As physicist Vera Kistiakowsky pointed out, "They have nothing to spin around." In the end, spin, as well, is only a metaphor.

Even the most helpful metaphors can lead to misunderstandings on a massive scale. I know whereof I speak. Like other science writers, I fell hook, line, and sinker for the metaphor that described the way matter gets its mass. The line we were handed was that particles acquire a certain sluggishness (or mass) as they slog through an invisible influence called the "Higgs field" like worms crawling through mud. A few years ago, physicists in Europe thought they saw a chunk of this mud when they hit it with fast-moving particles, and a piece flew off and left some tracks in their detectors—like mud in your eye.

The news hook was that this chip off the mud (properly known as the Higgs boson, or particle) would be an extremely important discovery because everybody wants to know the origin of mass. And Higgs, physicists were telling us, has the answer.

The sinker is that most of the ordinary stuff that makes up this book and its readers, as it turns out, doesn't get its mass from the Higgs field at all. It gets its mass from—hold on to your hats—nothing! That is to say, the protons and neutrons that make up the bulk of atoms are in turn composed of quarks. And quarks on their own weigh practically nothing.

The mass of the proton and neutron come, rather, from the frenetic twitching around of quarks trapped inside the bubbles in empty space, bouncing off the "walls."

As M.I.T. physicist Frank Wilczek puts it, "You start with massless particles and you get mass. That, to me, is much more satisfying and compelling [than finding the Higgs particle.]"

In fact, Wilczek likes this idea so much he has written in *Physics Today* that people who complain about gaining weight even though they eat practically nothing are, in essence, correct—and deserve our sympathy.

True, the Higgs is thought to endow elementary particles like quarks with mass. But matter, to most people, means sticks and stones and bodies and books. And when you talk about matter in that ordinary, everyday, sense, Higgs hardly enters the picture.

It isn't the fault of the metaphor, of course. It's the inability of our language to live up to the rich expectations of Nature. Any attempt to do her justice in everyday terms is ultimately doomed. As physicist Erwin Schrödinger, one of the founders of quantum mechanics, put it: "A complete satisfactory model of this type is not only practically inaccessible, but not even thinkable. Or, to be precise, we can, of course, think of it, but however we think it, it is wrong."

Literal Truth

It's the literal truth, we say, as if that "literal" conveyed an extra measure of authority.

Actually, literal meanings are frequently wrong, and often confusing.

An example is the "naked wife" virus that spread rapidly through cyberspace some time ago. The Department of Energy found it couldn't send out a warning about the virus because its prudish computer software interpreted "naked wife" literally—and censored the warning.

"Literal" means using a word in its exact sense, but is that how words are meant to be interpreted? As a kid, I used to fear that my mother would literally push me out of the car when she said she'd "drop me off" at a friend's house.

Anthropologist Vincent Crapanzano concludes in his book *Serving the Word: Literalism in America from the Pulpit to the Bench* that literalism is "a widespread characteristic of American thought," and a dangerous one at that. Literal thinking often leads to intolerance because it insists that only one meaning can be right—leaving no room for interpretation, ambivalence, or ambiguity.

Perhaps surprisingly, literal thinking has caused almost as many problems in science as it has in other realms of life. Consider, for example, the trouble physicists had trying to determine the nature of light when they were stuck with literal interpretations of the terms "wave" and "particle." Light has obvious wavelike properties, and also obvious particle-like properties; but it is clearly literally neither.

Eventually, physicists simply had to face the fact that light—like so many other physical phenomena—could not be strictly interpreted as either wave or particle. It required several interpretations, each dependent on context.

In fact, it's striking how often physicists got stuck in ruts because of literal thinking. They spent centuries trying to determine the exact physical properties of the so-called luminiferous ether—an all-but-undetectable substance that was thought to pervade all space. Taking the ether literally required believing that a substance could be at the same time as elastic as steel and as transparent as air. For a variety of reasons, physicists came to the conclusion that the ether didn't exist. But another very similar substance, known as the vacuum, is alive and well and at the center of physics.

Why does the vacuum persist while the ether was banished?

One reason is that everyone took the ether to mean one very specific thing, while the term "vacuum" evolved over time to embrace many phenomena, thus eluding a narrow, literal definition.

In the same way, one of Newton's greatest triumphs was his recognition at the outset that his phenomenally successful theory of gravity could not possibly literally work as his equations implied—by attracting objects toward each other at a distance through empty space. It took Einstein to figure out how it did work: Massive objects warped space itself, causing nearby objects to "fall" in their direction.

In the end, Einstein was able to revolutionize physics precisely because he freed words that other people interpreted literally from their narrow definitions—in particular, space, time, and speed.

Literalism is not just a disease of physics. In the late-eighteenth and nineteenth centuries, biologists noticed that developing embryos of mammals, including humans, developed tails, gill slits, and other fishy and reptilian features. They came to the conclusion that the embryo actually repeats the entire evolution of the species as it matures. This became known as the Biogenetic Law: "Ontogeny recapitulates phylogeny."

This "law" quickly fell from grace, however, when it became clear that embryos don't become fully formed fish or lizards on their way to becoming people or pigs. The idea was abandoned until the 1970s, when biologists realized that partial repetition of evolution is an important aspect of the process of embryonic development.

While the details were wrong, a central part of the theory was right. But literal thinking caused the baby to be thrown out with the bathwater.

Just because a theory turns out to be only partly true doesn't make it worthless. Paul Dirac did not discover "holes in nothing," as he thought. But he did discover what turned out to be antimatter. Genes do not determine behavior. But they certainly play a role.

Part of the power of science comes from the fact that new ideas are always built on the foundation of old ones. If experiments suggest a theory is wrong, scientists can either throw the theory out, or patch it up. Remarkably often, patching up works quite well—for example, putting it in a different context, or setting new kinds of limits.

It is people, not nature, who insist on absolutes.

Words, after all, are only what we make them. And remake them. So it isn't all that surprising that we sometimes can't agree on the proper meaning.

Or even what "mean" means, or "is" is.

Far-Out

Scientists ask people to believe the strangest things. They say that the universe exploded out of nothing and nowhere, and that 90 percent of the matter in it may be missing. They tell us that life blooms out of a simple solution of carbon and water, and that those selfsame carbon atoms, subtly rearranged, can sparkle as diamond, or dull the chimney as soot.

And then they look down their noses at people who believe in palm reading and astrology. Somehow, it's okay to believe in light-devouring black holes but not past lives; DNA but not ESP; ten-dimensional space, but not visits from aliens.

But how does one tell the difference between off-the-wall ideas accepted as "science" and off-the-wall beliefs that are just plain off the wall? What is the difference between science and pseudoscience anyway?

Ideas that can be tested by experiment or careful observation do not present much of a problem. Sooner or later, truth will out. But some truths are more elusive.

Haim Harari, a physicist who was also until recently president of the Weizmann Institute of Science in Israel, likes to tell the story of a recent UFO sighting in his native land.

The bright light was seen by hundreds of people, and soon after, almost as many explanations filled the radio waves as people called in with their theories—which ranged from stray bits of a satellite setting the sky on fire as they plunged into the atmosphere to aliens paying a call from a distant galaxy.

All sides agreed, said Harari, that "there is no solid proof that could rule out any of the possibilities."

But it didn't follow, said Harari, that all explanations were equally plausible. "Satellites exist, and little people coming from other galaxies are not known to exist."

Even though the satellite explanation could not be proved, it could be closely related to something known to be true. It is a small leap from real satellites orbiting Earth to a possible fragment of satellite burning up in our atmosphere. And so it is with even the most far-out ideas that scientists cook up. Ten-dimensional space may sound weird, but it's an extension of ideas that are tried and true. From the flat Euclidean space of your backyard to Einstein's four-dimensional space-time to ten-dimensional string theory is but a logical hop, skip, and a jump.

Compare that closely connected series of links with the idea that the positions of the stars or planets at birth can influence one's love life or financial future. It is not just that experimental tests have failed to confirm the reliability of these predictions; there is also no sensible way for planets or stars to exert that kind of influence on Earth. The gravitational pull of Venus or Mars on someone at birth would be overpowered even by the gravitational pull of the obstetrician walking around the delivery table.

Astrology, in other words, is not connected to anything known, while even black holes follow directly from what is known about gravity.

Of course, scientists themselves do not always agree on which wild ideas are worth pursuing. They do not entirely agree on the characteristics of black holes, the details of the Big Bang, or the plausibility of alien life in other solar systems.

"There obviously is no simple answer to the question [of

which wild ideas are worth pursuing]," said Caltech vice provost David Goodstein.

Far-flung flights of fancy called "thought experiments" are sometimes scientists' best friends. But at some point, even experiments carried out in the laboratory of the mind need to be tested by others. And as Goodstein points out, even the most abstract notions can be tested. Anyone with the requisite talents can do the calculations that revealed the link between gravity and black holes.

Making connections, after all, is what science is all about. The deep connections between electricity and magnetism remained unrecognized for many centuries, as did those between space and time, the fall of an apple and the orbit of Earth, and disease and microbes.

Science gains strength from these hidden roots. The scientist is like a gardener coming upon the thick network of underground tentacles linking the bushes and trees and flowers that seem so solitary on the surface.

Following the roots is one of the best ways to tell which wacky ideas are worth pursuing. The path isn't always straight or solid, but there is always a path—even if it sometimes takes you out on a limb.

Connected

It's a phenomenon well known to gardeners: Dig a hole almost anywhere, in the most seemingly barren ground, and instantly you find yourself enmeshed in a tangled network of interlocking roots.

On the surface, each flower and shrub may seem to inhabit its own patch of soil, but underneath, their long fingers secretly intertwine, weaving the world together.

Scientists and others who dig for the secrets of nature find much the same thing.

"When one tugs at a single thing in nature, he finds it hitched to the rest of the universe," wrote the naturalist John Muir.

The fall of the apple is connected to the orbit of the moon; the electric glow of lightning is linked to the attraction of magnets; space and time—like matter and energy—are aspects of each other, interchangeable as currencies; the atoms we're made of were conceived in stars.

During the past several decades, science has pushed its ability to connect to nearly unfathomable extremes. The tiny quantum mechanical fluctuations that tweaked the newborn universe, cosmologists tell us, grew up to be the large-scale geometry of filaments and galaxy clusters that drape the sky today.

The genetic recipes for apes and even insects and plants, biologists tell us, are written on the same spiral molecule as the genes of people—proving, if there was ever a doubt, the tightly knit nature of life's family tree.

The flapping of a butterfly's wing, mathematicians tell us, can stir up the air over Africa enough to set off an exponentially growing chain of global weather changes that causes a hurricane the next month in Miami. Light a candle in Hancock Park and send clouds over Kosovo. Pollute the skies in one state and rain acid on the neighbors next door.

"All things—linked are," wrote the English poet Francis Thompson. "That thou canst not stir a flower, / Without troubling of a star."

Recently, physicists even proved experimentally that subatomic particles can stay connected over vast expanses of space and time. In effect, once two particles interact, they become so entangled that they cease having separate identities. (If *Star Trek*–type teleporters ever beam us down on other planets, the technology will probably be based on this concept.)

What this means is that not even subatomic events are merely "local" anymore. "A monumental shift has taken place in our conception of things," writes Amherst physicist Arthur Zajonc in his book *Catching the Light: The Entwined History of Light and Mind.* "It goes by the humble name of 'nonlocality,' but within it is concealed a revolution in our thinking."

Scientists often follow these connections to deep truths about nature. Newton followed a similar link between the fall of the apple and the orbit of the moon to an understanding of gravity. Einstein connected falling with inertia and dug up his general theory of relativity—the idea that gravity results from the curvature of the fabric of space-time.

In the same way, Dmitri Mendeleyev followed connections between the behavior of various elements to the periodic table. Beneath all their apparent diversity, atoms, like species, can be grouped into families.

Everywhere scientists look in the universe, they see telling connections. If we look next door to our neighbor, Venus, we see a planetary hell hot enough to melt any spacecraft unlucky enough to descend near its surface. If we look toward our other neighbor, Mars, we see a dry-as-a-bone desert, shrouded in storms of thin pink dust.

And yet radar images of Venus reveal mountains and canyons eerily similar to those on Earth; Mars is marked with unmistakable evidence of ancient rivers and dry lake beds. What can we learn from these differences and similarities? Some geophysicists believe they see both past and future in these alternate sister worlds: Mars was once very similar to Earth; will Earth one day be similar to Venus?

Mathematics is the language of science in part because it helps reveal these hidden connections. The same equations describe the undulations of light, water, sound, tremors in the earth; the same mathematical pattern underlies the branching of trees, of blood vessels, of rivers.

The mathematical notion of symmetry is really the mathematics of similarities. It reveals that the subatomic building blocks of nature may be linked in previously unsuspected ways. Until recently, particles fell into two distinct families: roughly, those similar to light and those similar to what we more commonly know as "matter."

Now, however, physicists believe that every particle has a secret "super" partner lurking under the surface. Each member of each family finds its mirror image in the other world. If these hidden supersymmetric partners are found, it would mean that everything in nature—forces and matter alike—comes out of the same master equation.

The more things look different, the more they turn out to be the same.

Muriel Rukeyser reminds us in her poem "Islands" that swimmers sometimes see islands as "separate, like them." But no. "O for God's sake, they are connected, underneath."

Dubious Discoveries

The first time, it was a solo planet drifting around by itself in the vastness of space—the first ever seen untethered to a star.

The next time, it was "quark soup." Then it was hypothetical "dark matter" particles known as WIMPs.

Each time, scientists were so sure of what they saw that they called a news conference to announce the results to the world.

And each time, other scientists chimed in to protest: It just ain't so.

Why can't the scientists agree on what they saw?

To be sure, it's difficult to tell whether a dim point of light is a far-off star or a nearby planet. It's even harder to know what "quark soup" looks like, or to catch a WIMP in an ordinary trap.

Quark soup is an exotic state of matter that made up the whole universe just after the Big Bang—a boiling blend of subatomic particles called quarks and the gluons that hold them together. Quark soup can't exist in today's cool universe any more than hot water can exist for long in a freezer.

But then physicists in Europe thought they cooked up a soupçon of the stuff by smashing heavy particles at targets. Physicists on this side of the Atlantic weren't buying that a bit. Even if the soup did make a brief appearance, they argued, no one yet had tools to know what they saw.

As for WIMPs, suffice it to say that if dark matter were easy to see, someone would have claimed this discovery long ago. And WIMPs aren't named weakly interacting massive particles for nothing; they have so little contact with ordinary matter, in fact, that they pass right through most of it like ghosts.

Still, when an Italian group claimed to have found evidence for a dozen WIMPs a few months ago, seven major U.S. institutions put out a joint news release saying, in essence, No way. Better, more sensitive, detectors had ruled out the very WIMPs the Italians said they saw.

These on-again, off-again discoveries can make scientists sound like a bunch of puffed-up kids battling over bragging rights. One group claims: Yes, we did! The other insists: No, you didn't!

What's the public to think?

The truth is, it's not that easy to know when something has actually been discovered.

"You don't always get things right the first time," said physicist Roger Dixon of Fermi National Accelerator Laboratory in Batavia, Illinois, one of the institutions that put out the news release doubting the discovery of WIMPs. "The public is seeing the actual process at work. Sometimes that's confusing."

Indeed, a few months after its trumpeted appearance, the apparent solo planet turned out to be a dim star after all. And if the soup and the WIMPs evaporate under further scrutiny, as most physicists believe they will, the public won't necessarily have caught the scientists with their pants down. It will just have caught them in the act of doing what physicists normally do: making guesses based on the best available information, then holding their hunches up to the harsh light of experiment.

Theorists, after all, routinely ride their equations into dangerous territory, like the heart of the fiery Big Bang; they dream up all manner of exotic particles, spin out universes that breed like rabbits, and expand space and time into eleven dimensions. But these dreams come dressed in rules. The roller coaster of imagination comes equipped with built-in reality checks as brakes.

Like toddlers on tethers, the theorists are free to play in such dangerous ground precisely because at the end of the day, they're still firmly attached to the long arm of experiment. If their ideas are just hot air, eventually, the experimenters will bring them to the ground.

When experiments prove hard to do, of course, it's difficult to know when a theory has proved its worth. It's embarrassing to be caught yelling "fire" when there's nothing but a little bit of smoke. On the other hand, sometimes smoke is the only signal there is. But what if the smoke turns out to be ordinary dust kicked up by the experimenters' shoes?

The call is never simple. An everyday neutron can look like a WIMP, just as a man reaching for his wallet can look like a man reaching for a gun. Especially if you're expecting to see a gun. Or hoping to be the first human to glimpse the dark side of matter.

Experimenters are people, Dixon reminds us. "And some of them need to take credit [for perhaps premature discoveries] in order to survive."

The good news is, nature has the answer. The bad news is, sometimes we have to wait longer than we'd like to find out what she has in store.

Déjà Vu

We tend to think of scientists as forward-looking folk. But surprisingly often, they get ahead only by going backward.

Science progresses, in other words, by going back to the future.

Consider the current controversy over the mysterious "repulsive" force that appears to pervade empty space, pushing apart galaxies at the edges of the cosmos. Early last century, Einstein inserted a mathematical term for just such a force in his original theory of gravity. Later, when the force no longer seemed necessary to describe the universe, he took it out, calling it the greatest blunder of his life.

But scientists have been endowing empty space with powerful attributes since at least the ancient Greeks—who called it "quintessence," or the fifth essence. Newton and his contemporaries were convinced that empty space was filled with a "luminiferous ether" that carried waves of light like air carries waves of sound. Alas, this ether would have had to possess all sorts of impossible properties. It would have to be stiff, like Styrofoam, in order to vibrate fast, but planets and other objects would have to move through it without leaving behind holes.

Einstein figured out a way for light to move through empty space without the ether—and the ether got hidden away in the history books with other discredited ideas, such as alchemy. The idea of ether "was not really wrong," said physicist Jan Rafelski of the University of Arizona. "It was that we didn't need it any-more as a medium for light."

Now, it seems, the ether—or something like it—is making an encore, pushing the galaxies apart and eventually making the Milky Way a much more lonely place to live.

It's not only the ghostly energy of nothing that keeps com-ing back to haunt us. Even Einstein's ideas about relativity were not really new. Hundreds of years earlier, Galileo wrote at length about the illusions produced by moving frames of reference.

Someone trapped in the closed cabin of a steadily moving ship, he argued, could not tell whether he was moving—no matter what sort of experiments he performed: Drop a ball, watch fish swimming in a bowl, beam light across the cabin. A person standing on a nearby dock might say the ship was mov-ing, but the person in the ship could just as accurately say the ship was still and the dock was moving.

Indeed, the expression "everything is relative" describes Galileo's notions much better than Einstein's. Einstein's theo-ries deal more directly with what is *not* relative—for example, the speed of light.

And what about Copernicus? Wasn't he a forward-looking fellow? After all, he's credited with showing us that the sun, not Earth, was the center of the solar system.

But his motivations, according to Dartmouth physicist Marcelo Gleiser, were anything but forward-looking. Coperni-cus was trying to make the universe adjust to Plato's ancient re-quirements that all heavenly bodies orbit in perfect circles, and the Pythagoreans' idea that a fire burned at the center of the heavens.

"[Copernicus's] thought reflects a willingness to shake the very foundations of the cosmological ideas of his time, but only in order to reach farther back into the past," writes Gleiser in *The Dancing Universe.*

"He was, in short, a conservative revolutionary. He could never have guessed that, by looking so far back, he would be propelling civilization into the future."

Today, there's nothing more futuristic in physics than string theory—the idea that everything in the universe is composed of unimaginably tiny strings vibrating in ten-dimensional space.

But adding extra dimensions, it turns out, is also old hat in physics. Einstein introduced the world to four-dimensional space-time. Soon afterward, in 1919, Theodor Kaluza proposed that there might be a fifth dimension. In fact, he argued that both gravity and electromagnetism are higher dimensional ripples in the fabric of space-time.

And so it goes. "Progress does not always move in the forward direction," writes University of Warwick mathematician Ian Stewart in *From Here to Infinity.*

Luckily, there are a few old ideas that never came back. Alchemy, for example. Ancient chemists tried to transform lead into gold, but today's chemists know there's no way to turn one element into another. Or is there?

Truth be told, radioactive atoms do it all the time—naturally. And modern chemists working with supercomputers and the equations of quantum mechanics create entirely new kinds of materials from scratch.

In science, as in life, it's déjà vu all over again.

Credit

Like real life, science is inherently messy. Equipment shatters; experiments fail; results are ambiguous; data get dirty; equations defy solution; and causes are too complex to untangle.

But science is also messy because it's done by humans—among the messiest characters in the cosmos. Humans carry about on their shoulders not only big, rational brains, but also big, fragile egos; their hearts crave recognition, pine for fame.

"The product of science is knowledge," says the character of French eighteenth-century chemist Antoine Lavoisier in the new play *Oxygen*. "But the product of scientists is reputation."

So it's not a trivial matter when the credit for a discovery goes to the wrong person.

You'd think, perhaps, that giving credit where credit is due would be a simple matter. But it rarely is.

What if you stumble across a precious fossil, for example, but mistake it for a scratched-up rock? What if you know it's a fossil, but fail to publish? Or what if you figure out that there should be a treasure trove of precious fossils in your neighbor's backyard, but don't have the means to get at it?

A classic real-life example is the discovery of the cosmic microwave background—the afterglow of the Big Bang that still pervades space. The scientists who took home the Nobel Prize for this discovery had no idea what they'd found. They had picked up some stray noise in their radio antenna. For a while, they thought it was pigeon droppings.

Meanwhile, another group of astronomers had figured out that the Big Bang should have left behind still-cooling embers in the form of just such radiation. When they heard of the annoying "noise" raining down on the first group's radio antenna, they knew exactly what it was.

But in the end—to many scientists' consternation—the ones who stumbled upon it blindly got the prize.

Oxygen concerns one of the most curious controversies over credit in the history of science. In 1771 the Swedish apothecary Carl Wilhelm Scheele cooked up what he sometimes called "fire air" but never managed to publish his results. In 1774 the English minister Joseph Priestley created the same "vital air" and published his results within the same year.

Both Scheele and Priestley believed—as did most people at the time—that when things burned, a substance known as "phlogiston" was liberated. The gas that Scheele and Priestley created, they believed, was air sucked dry of phlogiston—or "dephlogisticated air."

It took the French tax collector Lavoisier to see that Scheele and Priestley had it backward: Things burn when something is taken from the air—and that something is oxygen. So while Scheele probably was the first to make oxygen, and Priestley the first to publish, Lavoisier was the first to understand what they had made.

Of course, there's more to it than that. There's the question of ethics, for example. Did Lavoisier steal his method of making oxygen from a letter Scheele wrote him? From information squeezed out of Priestley over dinner in Paris?

Where science is concerned, do morals matter? Is a Nobel Prize that goes to someone of questionable character somehow less noble?

These matters pertain outside the realm of science: Columbus may have "discovered" North America for the Europeans, for example, but his behavior was morally questionable. Besides, the Vikings probably got here first—only to find people already living here. And, of course, Columbus didn't know where he was going or recognize what he found when he got there.

Not coincidentally, the two authors of *Oxygen* are both winners of the Priestley Prize for chemistry. One, Roald Hoffmann, is also a Nobel laureate. (The other is Carl Djerassi.) So, they know whereof they speak: When it comes to winning prizes, self-promotion matters. So does whom you know. This is one reason why so many women scientists have not received the Nobel Prizes they almost certainly deserved (among them Lise Meitner for nuclear fission and Jocelyn Bell for pulsars).

And lest you think this is only ancient history, some juicy cases of credit are coming up just around the corner. Who, for example, will get the Nobel for the discovery of the Higgs boson when and if it appears?

The physicists at the European laboratory may or may not have had it within arm's reach before they were forced to shut down their accelerator to make way for a new one. The physicists at the Fermi National Accelerator Laboratory still have a shot at pinning it down. If they do, just who of the hundreds of experimentalists and theorists involved should be honored? And what of the CERN physicists who paved the way? To say nothing of Peter Higgs and other theorists who came up with the idea in the first place.

Nothing messy is foreign to the human heart.

Claims

So, whose big idea is it anyhow?

Big ideas are the stuff of fame and sometimes fortune for writers, Hollywood producers, corporations, and, frequently, scientists. And when big ideas are the coin of the realm, staking out claims can be a pivotal, sometimes highly contentious, issue.

As prospectors know, the gold goes to those who get there first. The second person to make a scientific discovery doesn't win the Nobel Prize.

Alas, establishing ownership of ideas is not an easy issue, as Steven Spielberg knows all too well. A few years ago, he was accused of stealing the idea for his film *Amistad,* based on an 1839 revolt aboard a Spanish slave ship, from author Barbara Chase-Riboud. Spielberg countered that Chase-Riboud had lifted portions of her 1989 novel, *Echo of Lions,* from a work titled *Black Mutiny.*

It all sounds eerily familiar.

Isaac Newton and Gottfried Leibniz battled over the origin of calculus, so much so that the British royal family tried to intervene. As historian Daniel Boorstin observed: "Eighteenth century Europe saw a vaudeville series of such bouts. Who had first demonstrated that water was not an element but a compound? Was it Cavendish, Watt or Lavoisier? Who was first to discover the vaccination against smallpox? Was it really Jenner—or Pearson or Rabout?"

Part of the problem is that ideas do not spring, full blown, like Minerva from the head of Zeus. They tend to evolve slowly—nurtured by extensive roots hidden deep underground. Suddenly, when the climate is right, up they sprout.

The question is: Who gets the credit? The one who produces the initial seed? The one who carries it to fertile ground? The one who raises it from seedling to mature plant? Or the person who happens along and notices what everyone else has ignored?

The ground was certainly fertile for thinking about energy transformation in the nineteenth century, when, within a ten-year span, twelve different people came up with the idea of the conservation of energy. Working in a vastly higher energy scale, two groups of physicists discovered the same subatomic particle in 1974, and for years, called it by different names—J and psi.

On the other hand, Aristarchus proved that Earth was round two thousand years before Columbus. But the world wasn't ready, and the idea didn't stick.

Of course, ideas by themselves are cheap. Every scientist (and science writer) gets scores of letters from people claiming to have solved the riddles of space, time, and gravity. Lots of poets wrote odes to trees, and any musical group could have written a song with the simple message: "I want to hold your hand." It's what you do with an idea that counts.

I was struck by the familiarity of ideas recently while listening to the *Odyssey* in my car, passing the time on the Santa Monica Freeway. The ancient world overflowed with the same spectrum of human emotions that fuels today's daytime soaps. And when it comes to splattered blood and spilled brains, today's filmmakers can't hold a candle to Homer.

Compelling ideas have staying power. They tend to resurface in newly fashioned—sometimes radically different—forms.

The real test is in the follow-through. One of the reasons Einstein gets so roundly labeled "genius" by his colleagues is that he carried his ideas to conclusions. He worked out the details that made the wildest ideas testable.

In these days of instantaneous electronic communications, scientists sometimes feel pressured to publicize ideas before they're ready—because they're afraid someone else will beat them to it, establishing priority. Then, however, they may get blamed for jumping the gun—especially if the conclusions turn out to be wrong.

On the other hand, there's little satisfaction in seeing a competitor take the limelight while you're carefully going over your figures, or working out a better method, or poking at your own work to see where there might be holes or soft spots.

Complicating the credit issue further, the person who solves the problem may not be the one who has the most productive idea. It may be the person who sees the problem in the first place.

As a physicist friend used to say: A smart scientist is one who looks at a problem and gets an idea for an answer. But a brilliant scientist sees a problem and gets an idea for a question.

Fail Safe

It's funny how the most intimate of personal dilemmas can mirror the most public scientific debates. For example: Should the United States spend $5 billion or more on a new particle accelerator? And if so, what kind?

The answer will turn, in part, on the same kind of considerations that go into deciding whether you should sign up for an experimental medical protocol, take a new job, make a pass at that girl.

How much risk are you willing to take and at what cost?

On the one hand, we are a society that puts the highest premium on safety: We want cars that can't crash, relationships that can't fail. At the same time, our communal cheer is: "Go for it!" As if these two approaches weren't mutually exclusive.

And so it is in science: Many big discoveries come from precise measurement and plodding observations: the beaks of finches, the motions of galaxies, the subtle loss of energy in particle collisions each in turn provided crucial evidence for evolution, dark matter, neutrinos.

Other discoveries, by contrast, involve expensive gambles: Space missions, decoding the human genome, most searches for subatomic particles.

Of course, big leaps also ultimately rest on the exacting collection and analysis of data. But there is a difference in both style and substance: Measure well what you think is out there versus go for broke and see what you can find that may or may *not* be out there.

Accelerator physics is inherently about careful measurement. It involves pumping up subatomic particles to high energies, crashing them together, and sifting through what comes out for clues to the fundamental nature of ourselves, the universe, and everything.

But even here, said UC Santa Barbara physicist Harry Nelson, "there are the control freaks versus the wonderment people."

Nelson was one of a dozen so-called High Minded Observers who attended a meeting in Snowmass, Colorado, on the future of particle physics. He was appointed by the organizers to keep his physics colleagues from getting too comfortable with conventional wisdom. So he was only doing his job when he bluntly told several hundred of them: "I feel we've underperformed. . . . We're not as good as we used to be."

Just why that is, he admitted, is a hard problem. But he thinks it has something to do with granting too much power to the control freaks. "There's something lacking we once had. It [once] felt much more exhilarating."

Particle physics hasn't had a completely unexpected discovery in a long time—something really off-the-wall, like the discovery almost a half-century ago that some particles prefer to be left-handed. "If you're in this for discovery, you feel a little disappointed," he said.

In some large part, that lack of surprise is testimony to the field's success. But it's also part of the pressure to "go where our theorist colleagues have already pointed us," said Nelson. "It feels like we're sort of monks copying things down from the books the theorists write for us."

One reason is, accelerators are expensive. Taxpayers (not to mention politicians) don't want to hear the scientists say: "We'd like to build a $5 billion accelerator, but we might not find anything new."

And, some things *can* be guaranteed from virtually any new machine: better measurements that lead to new theories or validation of old ones; a whole rich range of technological spin-offs. But major discoveries are another matter entirely.

Guaranteed discovery is an oxymoron. Asking for guarantees, said Jonathan Dorfan, director of the Stanford Linear Accelerator Center, "is a recipe for failure. If you're going to ask for guarantees, you're never going to reach the frontier."

Then again, "There is a big pressure to play it safe from our government," said physicist Jonathan Bagger, coleader of a panel charged with deciding the future of the field. "We're a very goal-oriented society." But going for a known goal can do more than preclude real discovery; being excessively goal-oriented takes away much of the adventure—the reason for discovering in the first place.

The late physicist Frank Oppenheimer liked to tell of a mistake he sometimes made as a young man exploring the Colorado mountains. The first time up a new path, he'd take his time, perhaps stumbling upon a beautiful meadow, or a waterfall, before reaching the grand vista from the top. But when he came up again with friends, he'd rush them in such a hurry to the summit that they never had a chance to see the meadow, the waterfall, or discover something of their own.

Oppenheimer compared this to the way physics is often taught in schools, but there's also a lesson for discoverers: You can't find anything unexpected if you always know where you're going. Sometimes the shortest distance from A to B isn't the most interesting or fruitful.

In the end, the biggest risk may well be not to take risks at all.

PART IV

"Political" Science

The Physics of Peace

There's no clearer example of why I gave up political science for "hard" science than the difficulty of understanding the terrorist acts of September 11, and the even greater difficulty of formulating an appropriate response.

Not that physics provides clear answers, but it does help shed light on the nature of the darkness, as well as the location of possible exits.

Here are a few ideas from physics that seem particularly relevant:

- The opposite of a deep truth can also be true: Waves can also be particles, and energy is another form of matter; two mutually exclusive things can be part of the same larger reality.

  Thus, hijackers who coldly employ air travelers as missiles can be religious people, described by neighbors as quiet, thoughtful, even kind.

  Conversely, we, the victims, aren't necessarily perfect or blame-free. As a recent immigrant from Belgrade so aptly put it, "If what attacked you is completely evil, it does not mean you are good."

- Every action has a reaction. Thus, the generous U.S. Marshall Plan after World War II created a Europe that was economically sound, educated, well-fed—and in the pinch, a strong ally of the United States. The heroism of New Yorkers rebounded back on them in tens of millions of dollars for victims and an abundance of donated blood.

At the same time, it's no secret that bin Laden himself is largely a reaction to anti-Soviet actions by the United States in Afghanistan that trained and armed fundamentalists and left the country devastated by war, desperately poor, and with nowhere else to turn.

- Everything is connected. Every electron to every other electron, every star to every other star. Muslims and Jews, bacteria and dinosaurs, plants and people—we all trace our lineage back to a common ancestor.

Thus, it's no surprise that citizens of sixty nations were killed in the bombings. Or that American flags waved in New York and elsewhere were held by people of every nationality and religion.

Alas, terrorists are also well connected, and targeting one won't root out the rest.

- Things look different in different reference frames. A friend speeding along at near light speed will appear to you to stay young almost forever, while in his own frame of reference will continue to get old at the normal rate. He will never be able to see things as you see them, or vice versa.

Thus our inability to "see" how someone can harbor enough hate to self-destruct and destroy so many innocent people, while in a different frame, some fundamentalist Muslims (and Christians) find it self-evident that these same innocents (being sinful) deserve destruction.

Reframing is a powerful tool. It can allow fanatic

Muslims—whose religion forbids suicide—to frame the act as "martyrdom," in much the same way as it allowed good Christians to take slaves by "framing" black people as property.

- Rules change in unfamiliar terrain, often in counterintuitive ways. In the tiny quantum world of the atom, particles can be here and there at the same time. On the scale of stars, space itself curves.

  Thus, just because something doesn't make any sense in our world doesn't mean it isn't true or doesn't need to be taken seriously. Russian veterans of the Afghan wars described trying to negotiate in a world where people don't care about material well-being. As one of them put it so poignantly:

  "Nothing we know works in their world."

There's more: Understanding the difficulties of separating signal from noise helps explain why clear warnings weren't heeded; inertia helps explain why sometimes it's difficult *not* to drop bombs.

Physics also teaches the importance of limits. That's why most physicists regard loose talk about a Theory of Everything as silly as boasts to "end all terrorism."

Of course, science not only fosters the understanding to get a handle on complexity, it also creates inventions that make us safer in an often dangerous world: antibiotics, indoor plumbing, airport security.

Perhaps someone can come up with a way to translate cultural beliefs from one frame of reference into another the same way that physics can translate from different moving frames.

Alas, we know less about the human mind than we do about

the edge of the universe. Still, in a psychological war that clearly requires psychological weapons, perhaps Madison Avenue's powers of persuasion could be brought into play. Poets and artists could be formidable weapons as well.

A physicist friend argued that more effective than bombs would be "goodie" drops containing food, medicine, books, CD players—promoting (as a reaction) goodwill and new allies. If a terrorist succeeds in making us terrible, of course, he wins.

Or as my physicist friend put it: "The worst thing a son of a bitch can do is turn you into a son of a bitch."

Default Lines

M y new computer was driving me nuts. Every time I tried to save a document, it would instead demand that I ask for "help" and refuse to budge until I went through the motions of posing some kind of question. As it turned out, the problem was easily fixed by altering the program's default settings.

Computer programs have minds of their own. That's why they're so useful. It only becomes a problem when you and the computer aren't of like minds and you don't realize you can fix the problem by changing the program's defaults.

Would that it were so easy to fix the defaults programmed into human thinking.

A familiar riddle illustrates the problem:

A man and his son are driving home from a baseball game when their car stalls on the railroad tracks. As a train rushes toward them, the father struggles desperately to start the car, but to no avail. The father is killed. The boy is critically injured. The ambulance arrives, and the boy is rushed to the hospital. In the OR, the surgeon takes one look at the boy and gasps: "I can't operate on this boy. He's my son."

If you haven't heard this before, you'll have some trouble figuring out what's going on here. But the solution is simple. The surgeon is the boy's mother. Alas, since our "default setting" for "surgeon" is male, almost everyone has a hard time coming up with the obvious solution.

"A default assumption is what holds true in what you might say is the 'simplest' or 'most natural' or 'most likely' possible model of whatever situation is under discussion," wrote Douglas Hofstadter some years ago in his *Scientific American* column, Metamagical Themas. "But the critical thing about default assumptions is that they are made automatically, not as a result of consideration and elimination."

We slip into default thinking completely unaware of what we're doing. We assume that the ground will be solid when we step on it (even though it sometimes isn't); we assume that the brakes will work on the car, that the words will come out of our mouths when we speak. Too often, people also learn to assume that blonds are dumb, Arabs are terrorists, that a group of black teenagers coming down the street means trouble.

Default settings are shortcuts that allow us to go through life without a thought as to what we are thinking. They are insidious because the default—in our minds—becomes the *only possible* scenario.

One species that put this to use is the cowbird, the female of which lays her eggs in the nests of other species. The bird that actually built the nest assumes that whatever hatches must be hers, feeding and raising the intruder as her own—even when the hatchling looks very different from the rest of her brood.

This is a fine strategy for cowbirds, but a very bad thing for the true fledglings who can get thrown out of the nest or starve as a result.

Scientists, too, mistake defaults for the only possible truth. They took it for granted that we couldn't know what stars were made of, couldn't fly to the moon (indeed, fly at all); they assumed that mountains are immovable and atoms indestructible. They took it for granted that space and time were dependable and solid; now we know they can slip, slide, stretch, warp, foam at the mouth.

Some argue that the default state of the human species is war. When we don't know what else to do, we fight. However, a recent study by UCLA psychologists throws this into doubt.

True, men under stress do tend to fall back on "fight or flight" as the first line of defense. Their default state is: hide, or hit someone. Women, however, cope with stress quite differently. They seek social contact or turn to nurturing—pick up the phone or tend the children. In fact, what principal investigator Shelley Taylor called this "tend and befriend" strategy for coping with stress seems characteristic of females of many species.

No one had seen this pattern before, Taylor said, because until recently, most studies focused on males—the universal default mode.

Given that the world is run mostly by men, however, it's going to take no small amount of imagination to think outside this particularly lethal box. Certainly, it's happened before. Against all odds, there is quasi democracy in Russia, quasi peace in Northern Ireland; Nelson Mandela prevailed in South Africa and even Nixon went to China.

Default is not the only option. In fact, Taylor and colleagues conclude that the "tend and befriend" strategy might well be responsible for the fact that women live on average seven and a half years longer than men.

An intriguing idea to consider for those of us who live so close to default lines.

(In)security

Physics is full of terms we assume we understand until we really stop to think about them: space and time, for example, or forces and particles.

So are the social and biological worlds.

Consider the word "security"—one we're hearing a lot lately.

We think we know what it means.

In Los Angeles, we like to feel secure against what some people call our four regular seasons: fire, flood, earthquake, and riot. Thus, we have fire extinguishers, sandbags, bottled water; some people have guns.

What makes you feel secure depends on who you are. It could be a good luck charm or a bad attitude, air bags or pepper spray, a big salary or a flu shot. For a government, it might mean putting a lid on dissent.

Most of the time, we think of a security blanket as a tangible object: a missile shield, an engagement ring, a lock on the door.

But it could just as easily be a talent—say, for running fast, or purring, or barking fiercely even though you're very small.

The variety of security systems that evolution has invented for personal defense is simply staggering. Porcupines are prickly, like cactus. Skunks stink. Bees sting. Snakes bite. Camels spit. Birds fly. Dogs run in packs.

Chameleons use camouflage.

On the species level, some solve the problem by proliferating wildly; they are everywhere at once, a vast network of individuals (like ants), so that if a million or so get done in, it really doesn't matter.

There are always a million more.

Evolution was certainly nudged along by this need for a clever array of defenses. To reproduce on dry land, amphibious creatures had to figure out a way to take a protected bit of watery environment with them. Presto, the egg. To get around and find food (or run away), they have to propel themselves on land. And so they got legs.

In fact, the evolution of life couldn't have happened if Earth hadn't produced its own protective magnetic shield to fend off electrically charged particles pelting it from the sun, and an ozone blanket to shade life from DNA-damaging ultraviolet radiation.

Needless to say, every security system has some holes in its armor (even Achilles had his heel). Surprisingly often, these weak spots are unintended consequences of other seemingly successful strategies.

For example, long, sharp teeth are fierce, but they break more easily than shorter, duller ones (ask a saber-toothed cat). The aerosols we used to protect coiffures poked a hole in the ozone layer. Large-scale use of antibiotics is just the thing to spur the evolution of microbes resistant to antibiotics, just as tough skin leads to predators with stronger jaws.

You have to choose your poison. Running in a pack won't work if you're too prickly or you stink. It helps to be nimble and quick, but then you can't carry around a hard, heavy shell. Then again, a heavy shell can be trouble if you roll over. Think turtles or SUVs.

Some of the recently proffered airport security systems are ironic in this respect. Take the idea of using sophisticated X-ray technologies that can strip-search people without taking off their clothes. To make ourselves secure, we strip ourselves naked.

Of course, sometimes the best security is just such transparency. Science, for example, relies on transparency to ensure that people don't cheat. Everyone has to be able to repeat an experiment. A free economy works best when there are no secret deals, nothing under the table. People feel secure when they think they know what's up, or at least what's going down.

This means they have to talk to each other honestly and freely. Monkeys and marmots and birds figured this out long before people came along. They're happy to squeal away, revealing their own position in order to warn others of imminent danger. But, like all mass communication, it only works if everyone is playing by the same rules—everyone squeals equally.

Happily, scientists tend to be squealers. In fact, in times of crisis, it is striking how often scientists of different nationalities keep right on talking even when their governments are at war. The transparent umbrella of science helps to make sure we don't overestimate—or underestimate—the enemy.

Scientific knowledge also leads to inventions that make us more secure: bulletproof vests, fireproof buildings, childproof containers. Not to mention weather satellites and Global Positioning Systems that give us due warning of severe weather or help us keep from getting lost.

But most important, science reminds us that real security comes mainly from good ideas, and the most successful defense mechanism that evolution ever invented is the brain.

And that, in turn, is what doesn't make sense about asking the public to clam up in the name of national security. As the naked emperor found out much too late, there's a lot to be said for an informed citizenry that squawks.

Feedback

For many years, I drove around with a bumper sticker on my car that read: STOP COMPLEMENTARY SCHISMOGENESIS! It probably did more to stop traffic than halt the escalating arguments to which it refers, but I liked the sentiment anyway. I stole the phrase from linguist Deborah Tannen, who uses the term to describe those little linguistic misfires that can so easily spiral out of control: Genesis, for beginning. Schismo, for rift. Complementary for actions and reactions that feed on each other. A relatively minor misunderstanding between men and women (or different nations) explodes into divorce (or war).

It is a typical example of what is commonly called a feedback loop: One thing pushes on another, which in turn pushes back on the first, which in turn causes the first to push back again, and so on and so forth, often with ever-increasing force.

Examples of such phenomena pervade the physical world as well. Consider stars. A star is born when a huge cloud of gas collapses under its own weight and the center heats up enough to ignite a nuclear fire. The fusion of hydrogen into helium releases enormous amounts of energy, which pushes back on gravity, halting the collapse.

This state of near equilibrium can last for a long time (our sun has been doing this for billions of years) but not forever. Eventually, the star runs out of hydrogen to burn. The core collapses, heating up this time enough to fuse helium into carbon and oxygen—a much more violent reaction. The star expands suddenly in what is known as a "helium flash."

Now the argument between the inward pull of gravity and the outward push of nuclear energy gets truly heated. And things speed up. If the star is big enough, the core collapses again, this time burning carbon and oxygen into heavier elements still. And so it goes, hotter and hotter, faster and faster, schismogenesis run amok.

By the time the star collapses entirely and explodes as a supernova, the whole process takes place in a matter of seconds.

The key to a feedback system is that whatever happens responds to what came before, over and over again. A predator gets so good at hunting down its prey that the prey population plummets. The predator goes hungry. The prey rebounds. Or not.

Feedback systems are tricky. For example, overfishing along the Pacific Northwest left killer whales with not enough to eat, so they started going after sea otters. Sea otters eat sea urchins. With fewer otters to gobble them up, the urchin population exploded. Urchins eat kelp, and so in turn gobbled up the kelp forests. Fish depend on kelp for both food and shelter, so fish lose out at both ends of the chain.

The complexity of feedback systems explains why climate, for example, is so difficult to understand. Global warming means the muggy air can hold more water vapor, which absorbs heat, warming the atmosphere even more. But more water vapor means more clouds, which could reflect the light of the sun back into space, cooling things down.

Alas, as feedback cycles speed up, things may happen too fast for effective intervention. Like stars, ecosystems and climate systems can rapidly collapse.

Not that all feedback leads to bad ends. Quite the contrary.

The collapse and explosion of stars seeds the universe with the elements that go into making, among other things, us.

Thermostats and flywheels are feedback systems. So is your body. If it gets too hot, it sweats, which cools you down. If it gets too cold, it shivers, which heats you up. If it runs out of fuel, it gets hungry, and you eat. If you get too full, you stop (chocolate, of course, doesn't count). Even the simple act of standing up requires constant feedback to keep you from falling over.

Science is a self-correcting system because it thrives on feedback from peers. Democracy is a solid form of government because of feedback from elections, polls, and a strong, free press.

We even invent social feedback systems to smooth our daily lives—like that little wave drivers give to other drivers kind enough to let them into a crowded lane, which encourages similar kindness down the line. Or so I like to think.

Feedback systems are destructive when they go too far, too fast. Like exit polling that influences elections, autopilots that overcorrect, politicians who lurch from one policy to another. Or nations that respond to attacks with ever-increasing counterattacks. Wouldn't it be nice if someone could invent a fuse for complementary schismogenesis that temporarily shuts down runaway feedback the way electrical fuses stop overheated circuits from melting down?

It would do a lot more good than my bumper sticker.

Neutrality

One of my students told me he was writing a paper about the efforts of the United States to maintain neutrality during World War I.

And I thought: Right. As if anything can remain neutral in our universe for long.

Strange to say, but the universe does not run well in neutral. Most of the time, the only way to stay in place is to keep moving. The status quo needs constant tending. Doing nothing is not an option—or perhaps I should say: Nothing doing!

Take something really simple. Say you are sitting on a chair, reading this book thinking there is not much going on (physically speaking), assuming that it doesn't require an exertion of force to stay put. You would, of course, be wrong.

If the ground beneath you weren't pushing back on your bottom with a force equal to that with which your bottom pushes on the Earth, you'd fall right through. In fact, gravity would cause the entire Earth to collapse if it weren't for the constant pressure of electrons in atoms elbowing each other out of the way.

It's often the way: Scratch the surface and "neutrality" becomes nothing but a cover for a sometimes precarious balance of opposing forces. As you sit on your chair, you don't think of yourself (or even the chair) as electrically charged. But you're electrically neutral only because the positively and negatively charged particles in your body balance each other out. If you were to pull them apart, the electricity in your body could probably power a good-sized city.

Or forget sitting, and stand up. What does it take to stand still? An enormous amount of continuous feedback from your brain to the balance sensors in your ears to the muscles in your legs simply to keep you from falling over. (Anyone who has ever tried to maintain a "neutral spine" while standing or doing exercise knows how hard this can be.)

Even then, you can't ever really stand (or for that matter lie) still. You're moving with the surface of a spinning Earth, which in turn is orbiting the sun, which in turn is strolling around the Milky Way, and so on and so forth. There is no standing still in our universe. You are always in motion relative to something, whether you can sense it or not.

Indeed, Einstein's equations describing gravity revealed—to everyone's surprise—that the universe couldn't remain static even if it wanted to. It has to expand or contract. Currently, it's expanding, but its future remains unclear.

Even nothing itself isn't neutral. The vacuum of empty space jitters with continual uncertainty; particles of matter are coming into existence all the time, exactly balanced by an equal number of particles of antimatter; indeed, this energy of emptiness may well account for most of the energy in the universe— even though in terms of matter, it adds up to nothing.

Scientists know that it's a lot easier to "neutralize" a problem than to eliminate it completely. For example, astronomers would see the stars a lot more clearly if they could eliminate distortions caused by turbulence in the air—in effect, take the twinkle out of starlight. But it's hard to get the atmosphere to stand still, so instead, a technique called "adaptive optics" keeps track of the wiggles in the atmosphere, then exactly counters the wiggles to make them go away. It's like adding minus one to one to make zero. Thousands of times per second.

At some level, even smart politicians know there's no such thing as neutrality. That's why we have checks and balances built into democracy, economic policies, and international agreements, so that everything has a way of pushing back on everything else. It's a lot of work maintaining stable balance, but really, there's no alternative. Equilibrium is always dynamic.

In biological systems, neutrality is death. An organism that doesn't constantly put energy to use fighting disintegration dies. The Second Law of Thermodynamics guarantees it. Disorder is the natural order of things and if you don't battle it constantly, things fall apart. Shoelaces come untied. Rooms get messy. Teeth decay. Waists spread. Food rots. Tires wear. Everything from schools to relationships deteriorates.

Even mountains, sooner or later, get worn down, and if the energy of radioactive decay didn't continually work to uplift them, the surface of Earth would be flat.

Increasing disorder (quantified as entropy) is what makes the arrow of time point only one way; it's a phenomenon so familiar that everyone understands intuitively that the universe can't run in reverse.

Sometimes we need to remind ourselves that it doesn't run in neutral either.

Dreamers

Particle physicists are depressed. So, too, are planetary scientists. A recent string of science budgets streaming from Washington threatens to defer their dreams of exploring both inner and outer space.

"So what?" you might well ask.

The research they do is admittedly speculative, esoteric, impractical. It costs millions—sometimes billions—of dollars. Dollars that are urgently needed for the government's wars on terrorism, cancer, drugs.

Who needs dreamers in a time of tangible terror? Far-out research when so many down-to-earth matters need attending?

In the current climate, it's perhaps not surprising that missions to Pluto and Jupiter's moon Europa have been put on hold. Even though postponing the trip to Pluto means giving up a chance to visit the last unexplored planet in our solar system for at least a hundred years. Even though strong evidence suggests that Europa holds an ocean under its ice—making it a good bet for finding the first sign of extraterrestrial life.

It's also not unexpected that funding would be steadily reduced for particle physics, that branch of science that unravels the innards of atoms—the ultimate in navel gazing. In February 2002, director of the U.S. Office of Science and Technology Policy John Marburger—former director of a particle physics lab himself—called the frontiers of astronomy and particles both "remote hinterlands" that were "no longer very relevant to human affairs."

The folks at Fermi National Accelerator were aghast. "Remote hinterlands?" exclaimed Judy Jackson in the in-house paper *FermiNews*. "[And] these words come from a *friend* of particle physics."

Reflecting Marburger's attitudes, the 2003 science budgets are all about "missiles and medicine," as insiders are calling them—with huge increases going to support research in the Department of Defense and the National Institutes of Health.

But here's the odd thing: It's the dreamers in their hinterlands who often come up with the most practical inventions of all—those most relevant to human affairs.

Take that terrorist with the little black mustache who committed mass genocide and came within arm's reach of taking over the free world not so long ago. He might have succeeded had he managed to develop his atomic bomb. One reason he didn't was that most of the German and Austrian "dreamers" fled to the United States.

Just look at the physicists behind the U.S. atom bomb: They were a bunch of fuzzy-headed, highfalutin intellectuals—engaged in spectacularly impractical research. Oppenheimer was a theorist who explored black holes. Einstein came up with the famous theory that revealed you could get energy from matter ($E = mc^2$) while daydreaming about what it would be like to ride along on a light beam.

So much for missiles. What about medicine? One can't well begrudge funding medical research that ultimately helps everyone (well, everyone who can afford to pay). But where do PET scans and magnetic resonance imaging and laser surgery come from?

You guessed it—dreamers! Magnetic resonance sprang right from the innards of atoms, lasers from quantum mechanics; PET scans rely on antimatter. No one had practical technology in mind when these discoveries were made.

It would be unfair to stop with fuzzy-headed physicists. There are fuzzy-headed biologists doing the same sorts of work. Tramping around the jungles looking for new species no one will ever keep as a pet or use for carrying loads or eat. Most of their discoveries are tiny insects and silly plants.

Surprisingly often these finds from the "remote hinter-lands" of biology lead directly to new medicines, or new understanding of disease.

This particular lesson was brought home in a big way recently when Japanese computer scientists left their U.S. counterparts in their dust. According to the *New York Times,* the Japanese have produced a computer more powerful than the twenty fastest U.S. computers combined. They developed it in the process of trying to understand climate change, modeling weather and earthquakes and global warming.

The United States, meanwhile, has been developing computer power primarily to model weapons. And look what happened.

"These guys are blowing us out of the water," one U.S. scientist said. Another called the situation "Computenik"—referring to the astonished U.S. response when the Soviets sent the first artificial satellite, *Sputnik,* into space in the 1950s.

(Back then, tremendous resources suddenly poured into science education. Not this time. Instead, millions of dollars previously earmarked for math and science education have recently been stripped from an education bill.)

It is part of our Puritan ethic that work should have a clear purpose. Journeys should have destinations; efforts clear ends. There's no room for leisurely wandering, pointless wondering: What would happen if...? What's out there? In here?

Alas, this ethic simply doesn't serve science, where seemingly aimless exploration is often the most fruitful of all.

Puritan ethic or no, there's nothing idle about scientific curiosity.

Natural Law

The Laws of Nature. The term has such a clear, comforting sound. We like to think that the universe is governed by the strict rule of law: consistent, predictable, without exception.

Laws of nature seem as indelible as the Constitution, but far more effective. You can avoid taxes, but you can't defy gravity; you can drive faster than the speed limit, but nothing can exceed the speed of light. Of course, you won't go to jail for ignoring a law of nature. On the contrary, if you broke one, you'd probably win a Nobel Prize.

The stereotypical view of nature's strictures is well summarized in a popular T-shirt that shows Einstein as traffic cop, holding a sign: SPEED LIMIT OF THE UNIVERSE: 186,000 MILES PER SECOND. IT'S THE LAW!

Our current national obsession with "zero tolerance" before the law probably stems at least in part from the belief that "zero tolerance" is nature's way. Legal laws, like natural laws, should apply uniformly in all circumstances without exception or exemption.

Alas, this belief is mistaken. Scientists have always been in the business of breaking (or at least stretching) current law. The Ten Commandments may be set in stone, but the laws of nature are not. Good scientists study the laws, but great scientists discover their weaknesses.

"One of the ways of stopping science would be only to do experiments in the region where you know the law," said the late Caltech physicist Richard Feynman in his classic work on the subject, *The Character of Physical Law.*

Consider a simple, universal law—like the gravitational attraction of every bit of matter in the universe for every other bit of matter. Newton's laws of gravity set physics on such firm footing that astronomers were able to deduce the existence of Neptune from a slight irregularity in the orbit of Uranus; centuries later, the same laws of gravity set men on course to the moon.

But Einstein showed that Newton's laws were incomplete. The old laws broke down in subtle ways, or in extreme circumstances—in the fierce gravitational field of a star, for example.

The same story is repeated again and again throughout the history of science. Ultimately, all laws are tentative. "We are never definitely right," said Feynman. "We can only be sure we are wrong."

What, then, is a law of nature? According to Feynman, it's an observed pattern or rhythm to the natural world. Drop an apple (or a desk), and it always falls. Sometimes, these patterns are "not apparent to the eye, but only to the eye of analysis," said Feynman. If you drop a piece of tissue paper, it does not appear to fall in the same way as a rock. But the kinship of the pattern is revealed once you strip away air friction.

A law of nature, in other words, is a steady relationship, something that always happens—or at least as far as we know.

But the most interesting science happens where the laws break down. That's one reason scientists like to push the laws to the breaking point, to see where they crack.

Currently, most physicists believe that Einstein's laws of gravity break down in the extreme space and time warps inside black holes or at the origin of the universe. According to Cambridge University physicist Stephen Hawking, Einstein's gravity predicts its own downfall.

Physicists plumb other cracks in the laws for signs of new and exciting physics. For example, for unknown reasons, the universe is made of matter, but not antimatter—even though known laws predict that it should contain equal amounts of both. "The laws of physics are not quite the same for particles and antiparticles," says Hawking. Finding out why is a major thrust of massive research efforts on several continents.

Despite persistent efforts to come up with a "theory of everything," explaining everything is not what science does well. "Physics does not endeavor to explain nature," wrote the late physicist Eugene Wigner. "In fact, the great success of physics is due to a restriction of its objectives."

Limiting objectives allows physicists to learn which laws apply in which circumstances and which don't. And just as laws against weapons in school may not apply to eight-year-olds carrying penknives in their lunch boxes, laws of gravity may not apply in the extreme time warps of black holes.

When the laws lead to absurd or nonsensical answers, physicists know it's time to look for new ones. Sometimes, exceptions rule.

Popular Science

I was sitting in the bleachers at the Hollywood Bowl listening to Gershwin's *Rhapsody in Blue,* when an irritating thought shattered the serenity of the evening. "I can't listen to that anymore without thinking of United Airlines," I whined to a friend. Advertisers shouldn't be allowed to hijack great music to sell things.

My friend, however—someone highly placed in classical and opera circles—quickly put me in my place. "I think it's great. It exposes people to classical music."

Ouch.

Truth be told, I've responded the same way when people complained about "popularizing" science. In fact, my ill-considered complaint about using Gershwin to get people flying reminded me of the interviewer who once asked me if I wasn't uneasy about making math and physics accessible to the masses.

"Isn't that cheating?" he asked. "Don't you think it's wrong for people to want to learn about math without doing all the work?"

It's a curious point of view. After all, we're a nation of voyeurs. People who never get off the couch spend long hours absorbed in basketball, football, hockey, golf. We take vicarious pleasure in everything from the seedy secrets of Hollywood stars to the sex lives of presidents. We drive cars and surf the Internet, even though few of us can change a tire or create a Web site. We listen to Gershwin, even though we can't play a note.

Is participation without mastery cheating? Is enjoyment without "work" legit?

Art, like science, seems to be of two minds on the subject. On the one hand, outreach is the order of the day. If more people aren't brought into the tent, classical music, like physics, may wither for lack of enthusiastic young supporters.

On the other hand, experts worry that making their subject accessible to the hoi polloi strips it of its depth, tarnishes its delicate beauty. Like substituting plastic flowers for the real thing, taking Latin out of the liturgy, or teaching physics without math.

In science, these worries about the perils of popularization have surfaced as the so-called Carl Sagan syndrome. Sagan's classic TV series *Cosmos,* along with his many popular books, brought forefront physics and astronomy into millions of minds and hearts. He not only allowed people to cheat, he encouraged it. And by most accounts, he did an excellent job.

Yet Sagan paid a steep price. He was dismissed by many colleagues as a mere popularizer and shut out of the prestigious National Academy of Sciences.

Other scientists fear the same will happen to them. Rather than take the chance of getting crucified for playing to the crowd, they'd rather, as Marie Antoinette might have put it: "Let them eat calculus!"

Of course, it's rather silly to think experts can keep either science or art locked up as private treasures, privy only to those with the proper credentials, safe from the great unwashed. Novelists, playwrights, composers, and painters routinely appropriate ideas from modern science. Tom Stoppard based his hit play *Arcadia* on chaos theory. Brazilian composer and singer Gilberto Gil incorporates quantum theory into his work.

Yes, there is a danger here. The ideas that appeal to talents like Stoppard and Gil are also fodder for outright flakes, or simply slipshod treatment. When science slides down this sometimes slippery slope, it can lose the very rigor that makes it solid ground from which to explore everything from alternate universes to the physical basis of consciousness.

And yet, the alternative seems equally unacceptable. The idea that amateurs shouldn't dabble in science has led to a terrifying situation, according to Harvard physicist and historian Gerald Holton. It's not only that the man and woman in the street know almost nothing about science, he writes in his collection, *Einstein, History, and Other Passions.* Almost all of our intellectual and political leaders are equally ignorant.

"All too many find themselves abandoned in a universe which seems a puzzle on either the factual or the philosophical level," he says in the book. "Of all the effects of the separation of culture and scientific knowledge, this feeling of bewilderment and basic homelessness is the most terrifying."

Perhaps all this seems a tad self-serving, coming, as it does, from a professional peeping Thomasina like myself.

But if we can enjoy spectator sports, why not spectator science and art?

Perhaps people will become so enthralled by what they see on the surface that they'll want to dive in more deeply.

And next time someone hears the United Airlines theme song, they'll think of Gershwin. Instead of vice versa.

The Science of Art

What's art got to do with it?

A lot more than people generally think.

To educators fighting over school budgets, art and music frequently are viewed as frills that drain funds from more serious subjects like math and science. But scientists and mathematicians know different. In fact, they often rely on aesthetics to guide their research, filter their perceptions, and help them visualize patterns in the sometimes impenetrable chaos of data.

That's why recent efforts in some areas to put the arts back into the schools is such good news for science education. Among the children who will benefit most are the future scientists and mathematicians—and the people who come to use their discoveries and inventions. Artistic training can sometimes play a critical role in scientific success.

Scientists have long said that the best of their breed are artistically inclined. Most everyone has seen photos of Einstein with his violin and physicist Richard Feynman with his bongos. I've sat next to physicist Frank Wilczek while he played silent Bach piano concertos on his knees during professional talks. Nobel Prize–winning chemist Roald Hoffmann writes highly praised poetry (only sometimes about molecules).

Put four mathematicians in a room, the old saying goes, and you're sure to have a string quartet.

In fact, artistically inclined scientists tend to win more awards than their less diversified colleagues, according to several studies.

Michigan State University physiologist Robert Root-Bernstein and his psychologist mother, Maurine Bernstein, found that most Nobel Prize winners and members of the National Academy of Sciences had arts-related hobbies.

"Their less successful colleagues did not share either their arts interests or their arts-related thinking skills," the authors concluded.

This finding, replicated in several similar studies, seems a logical extension of other research conducted at UC Irvine suggesting that exposure to music actually enhances intellectual ability. Not only does listening to Mozart improve test performance (at least temporarily), preschoolers who play piano do better at science and math than their counterparts who don't.

Why should this be so? Why should painting or playing piano or writing poetry have anything to do with math or science? One obvious reason is that scientists, like artists, must learn to pay close attention—both to detail and to the broader context. Scientists, like artists, are people who notice things. They not only see things that other people often ignore, they also see the frequently hidden links among disparate aspects of reality.

Scientists and engineers, says Root-Bernstein, "must learn to observe as acutely as artists and to visualize things in their minds as concretely. They must learn to recognize and invent patterns like composers or poets ... and play their high-tech instruments with the same virtuosity as musical performers."

Another art-science connection may lie in the relationship between our hands and our brains. A book by California neurologist Frank Wilson, *The Hand: How Its Use Shapes the Brain, Language, and Human Culture,* argues that people who use their hands are privy to a way of knowing about the world inaccessible to those not schooled in manual arts.

Speaking on National Public Radio's *All Things Considered,* Wilson told of a car mechanic who got a call from a vice president at a big computer company, complaining that his MIT-educated engineers couldn't solve problems as well as the older engineers at the company. It turned out, Wilson said, that 70 percent of older engineers fixed their cars, and 20 percent had some experience with wrenches. Of the young hotshots, none had ever held a wrench. As a result, they weren't as good at understanding complex systems.

The hand's knowledge about the world, according to Wilson, actually teaches the brain new tricks. The hand's touching, exploring, and manipulating can rewire the brain's neural circuitry.

Finally, logic alone is sometimes insufficient to solve really complex problems. Even Einstein said that imagination was more important to a scientist than knowledge. Physicists on the forefront of discovery often talk about being guided by "smell" or instinct. They talk about the "aesthetic" appeal of ideas.

According to French mathematician Henri Poincaré, aesthetics was "a delicate sieve" that helped scientists sort through the confusion of facts and theories. The physicist P. A. M. Dirac observed: "It is more important to have beauty in one's equations than to have them fit experiments."

Painting, piano playing, and poetry help put things in context, sharpen details, hone observations. They sort the essential from the peripheral, forge connections, find patterns, and discover new ways of seeing familiar things.

These are exactly the tools any good scientist needs.

Small Potatos (sic)

People should stop picking on certain politicians just because they have poor syntax. Far more troubling are their problems with math.

Consider the simple matter of multiplying two times two. Doubling is as simple as it gets, and it's something that happens to everything growing at a given rate. If your money is earning 7 percent interest, the "doubling time" is ten years. If the rate is lower, the doubling takes longer, but sooner or later, it doubles just the same.

The same simple math is behind everything from population growth to increasing rates of energy consumption. And doubling adds up astonishingly fast.

My favorite dramatization of this ever-present phenomenon comes from University of Colorado physicist Al Bartlett. He tells the tale of two bacteria that take up residence in a Coke bottle at 11 A.M. (Call them Adam and Eve.) They beget and beget, doubling numbers once a minute. At noon, their bottle is full.

What time would it be, Bartlett asks, when the most far-sighted politicians in Bacterialand notice that they are running out of room? The answer is 11:59, when the bottle is still half empty. (One doubling time away from full.)

Suppose, Bartlett says, that the bacteria decide to drill off-shore for new Coke bottles and turn up three pristine, never-before-inhabited bottles. How long before they run out of space again? You got it: two more minutes.

Play with the numbers how you will, no matter how many new resources you discover, so long as the rate at which you are using the resource continues to grow, you run out sooner than you think.

(And, as Bartlett points out, there's geometry involved here as well. If we lived on a flat Earth that extended infinitely in all directions, we might never run out of energy or oil. But a spherical Earth has a finite surface area. For this reason, Bartlett likes to call those who think we can grow forever the new Flat Earth Society.)

It's not just misunderstandings about multiplication and geometry that make bad public policy. It's probability as well. For example, cigarette makers long argued successfully that, since you can't predict which smoker will die when, it's merely a matter of *probability* that someone will get lung cancer or heart disease from smoking. Similarly, gun supporters argue that having a large number of guns in people's hands only increases the *probability* of murder: Guns don't kill; people do.

Both these arguments rest on the assumption that probability isn't a cause. But any Las Vegas casino operator can tell you different: The only reason that seven comes up more often on a throw of two dice than any other number is that it's the most probable combination. Then again, it's probability that made Humpty Dumpty fall into a thousand pieces.

Many important insights have also emerged from a lesser-known branch of mathematics known as "game theory." True, an understanding of game theory might help you beat the market, but it's also uncovered some curious truths about, for example, the relative merits of competition and cooperation. In a series of now-classic studies, competing computer programs (computers can "play" faster than people) tried out various strategies for long-term survival. Those that put an emphasis on cooperation fared much better in the long run than those that employed competitive, confrontational tactics.

Game theory may give some insight, in other words, into the reasons arms agreements have led to forty years of nuclear peace—and why politicians should do the math before altering that balance.

It's probably wishful thinking to suppose that an administration that took so long to hire a science adviser will trouble itself with a math tutor.

Still, at the expense of alienating English teachers, it's a lot more important for politicians to be able to multiply two times two than tell a potato from a potatoe—or know an inalienable right from an uninalienable one.

A few lessons in math could easily reveal the folly of thinking we can solve our energy problems by searching for new Coke bottles in the Arctic National Wildlife Refuge, for example.

The only lasting solution—at least so long as we live on a spherical planet—is to slow the rate at which we use up those bottles.

It's as simple as two times two.

The Geometry of Fairness

The last place you'd think to find insights from Einstein is the debate over affirmative action. And yet, there's a sense in which the whole thing is a fairly simple problem in geometry.

The underlying question here is: What's the shape of the playing field?

If the playing field is tilted in favor of women and minorities, then obviously affirmative action is unneeded and unfair to white guys; if the playing field is tilted against women and minorities, then affirmative action just as obviously is needed.

Alas, few people stop to think about the shape of the stage on which we play out our lives, mainly because it's normally invisible. But ever since Einstein refashioned the way we think about space and time, it's become a real factor in every physical equation.

To get a sense of how the geometry of the playing field can change things, consider the following riddle. You are at some unknown location on Earth. You walk one mile south, make a ninety-degree turn, and walk one mile east, then walk one mile north. You are back at your starting point.

What color are the bears?

The answer is white, because you're at the North Pole.*

Of course, anyone who took geometry knows you can't

---

*This is not literally true, of course; there are no bears at the North Pole.

make three right-angle turns on a flat surface and get back to your starting point. But if the surface is curved—like the surface of the Earth—you can do all sorts of things your geometry teacher never taught you.

For example, most people learned in geometry that two parallel lines never meet. Again, this is true enough for space that's flat and two-dimensional like a piece of paper. But two lines of longitude that are parallel at the equator meet at the poles.

The shape of the background, in other words, makes a huge difference in how things work, whether we're aware of it or not. And it affects a great deal more than lines and angles; it can also determine how physical forces act.

According to Einstein's relativity, for example, gravity is really the result of the curvature of four-dimensional space-time (a joining of space and time that creates the backdrop for our universe). So it's the geometry of the unseen background that determines what rises and what falls on its face.

How can you measure the shape of something you can't even see? Surprisingly, perhaps, it is doable. Take the curvature of space-time. Even though you can't see it directly, you can measure its warp by observing the way it bends light from distant galaxies.

You can even determine the shape of the universe at large by measuring the way its contours affect light reaching us from the farthest shores imaginable—the afterglow of the Big Bang. If the light makes a triangle whose three angles add up to more than 180 degrees, you know you're on a surface that curves in on itself—like the surface of the Earth. And if it makes a triangle whose angles add up to less than 180 degrees, you know you're on a surface that curves outward—like a saddle.

Scientists use these tools to get a grip on such unwieldy

matters as the ultimate fate of the universe. If the universe curls in on itself, then it will eventually collapse into a single point; if the universe curves outward like a saddle, then it could keep expanding indefinitely.

To be sure, debating affirmative action requires measuring devices different from rulers and protractors. Still, if we can measure the shape of the universe at large, it shouldn't be all that difficult to come up with a set of tools for figuring out the shape of the economic and social playing field here on Earth. One way to start is to look at the patterns of who tends to rise—and who consistently falls. If it's always the same players, one begins to suspect that the shape of the playing field needs fixing.

Faster

Los Angeles high-school teacher Guillermo Mendieta was going on a hunger strike.

Not over salary, or benefits, or working hours. He was putting his passions where his mouth is over—of all things—mathematics, and the way it's taught in L.A. schools.

Specifically, Mendieta was trying to stop the Los Angeles Unified School District from giving up on a decade of reforms in math teaching. He and others feel these reforms have brought the power and pleasures of math to students formerly written off as hopelessly math-inept.

Many of these students are black or Latino. The rest are almost everyone else who didn't breeze through math in high school: Allergic to algebra. Terrified of trig. Catatonic over calculus.

These widespread aversions are side effects, many believe, of the drill method of math teaching, which is about as likeable as the drill method of dentistry. As antidotes, reformers brought in everything from hands-on activities to art and games.

They added alternatives to the single-minded math track that forces students to march inexorably from algebra to geometry to pre-calculus, built bridges between various branches of math, and even blazed trails from math to science, humanities, literature.

Back-to-basics advocates argue that while reforms might have made math more appetizing, they don't provide the kind of substance that leads to real competence. Appealing, perhaps, but junk food. This lack of competence, they say, shows up in the ever-dismal test scores of U.S. math students—particularly those in L.A.

To be sure, tests are invaluable as a tool to measure what students know and where they stand. But tests can't tell you everything.

Consider that famous turn-of-the-century math student, the horse Clever Hans. When Hans's owner asked the horse to add, say, 3 plus 5, Hans pawed out the right answer. Hans couldn't do sums, of course, but he could read his owner's silent cues (perhaps no less of a feat).

Lest you think Clever Hans is a quaint historical oddity, the same phenomenon turns up today in the most amazing places—Harvard, for example. A documentary produced some years ago by Harvard and the Smithsonian Institution, called "A Private Universe," filmed newly minted Harvard graduates as they struggled to answer the question: Why is it warmer in summer than in winter? Most got it wrong.

All these students, one presumes, were above-average test-takers. Some had even studied astronomy. But like Hans, they had answers without understanding.

Math tests, in particular, tend to be timed—which rewards students who can solve problems quickly. This is a good, useful, and sometimes important skill. But not always. "To most professional mathematicians, the focus on speed is crazy," says mathematician Keith Devlin, author of *The Math Gene* and a half-dozen other popular math books. "Most of the good mathematicians I know are very slow. All the tests really discover is whether you can do something fast."

Devlin and others worry that judging the success or failure of math teaching by test scores—especially timed tests—discourages students more than it helps teachers evaluate what they need.

His own daughter was left back in high-school math for poor performance on tests. Given the time she needed in college, she went on to make straight As. When Devlin wrote about his daughter in *Focus,* a publication of the Mathematical Association of America, he received hundreds of passionate letters recalling similar experiences.

Passion, of course, is something no test measures. And it's passion that propelled Guillermo Mendieta to go without food in order to make his singular appeal. He knows how it feels to be starved for the confidence that comes from being at home in the world of numbers, the pain of exclusion that dooms most children to a life of feeling inadequate—okay, let's just say it: stupid—in math.

To go through life feeling muddled by math, Mendieta understands, can be as impoverishing as going through life hungry for food.

So before the school district junks math reform, it needs to think hard about the way it measures success. The paths to mathematical literacy are many and varied. Depriving kids of the good feeling that comes from knowing a thing or two about numbers would be far worse than letting them eat a little cake.

Unnatural

Who can resist the lush lure of the natural? The splendor of the grass, a roll in the hay, the almost sinful softness of rose petals?

There's something seductive about the heady aroma of the farmers market on Sunday, the rough-and-tumble rowdiness of the dogs playing in the park. Nothing will sell a house faster than fine wood floors, fireplaces, and ocean views.

It's only natural that we cringe at the artificial: It goes against the grain. So it's no real wonder that many people are feeling uncomfortable about the proliferation of genetically engineered crops. Protests sprung up all over Europe, and the French have taken to calling these products "Frankenfood."

It makes us understandably uncomfortable to eat corn "enhanced" with the genes from bacteria—even if those genes protect the corn from insects.

And yet, we have to remind ourselves: There's nothing "natural" about a dahlia, or a house cat, not to mention the melons and corn and tomatoes that call to us from the farmers' stalls—even the completely "organic" ones. Human manipulation of genetic traits by crossbreeding is nearly as old as civilization. White peaches aren't "natural" any more than the "natural" cereal in my kitchen cupboard grows on trees. There's nothing natural about bread, or wine, or beagles.

In fact, there's nothing "natural" about ourselves. In effect, humans and our ancestors evolved only because previous microbial inhabitants of Earth polluted the planet with a "poison" called oxygen, on which we happily thrive.

If it all seems confusing, it is. And not only in the world of the living. For example, most people would say that plastic is decidedly "unnatural." But as chemist Roald Hoffmann points out, most plastics are made from petroleum products, and petroleum is a product of long-dead plant matter that brewed for millions of years in the bowels of the earth.

So plastic did, in a sense, grow on trees.

Even physics has a long history of confusion over "natural" laws. For example, Aristotle thought it was "natural" that light things such as clouds should rise in the sky, while "heavy" things such as rocks should sink. He thought it was "natural" that heavenly bodies should move in circles.

Later, physicists realized that the force of gravity pulled rocks to the ground and planets into orbits. Left to their own "natural" devices, they would simply float about in space.

Later still, Einstein discovered that falling objects and orbiting planets were, after all, simply following their "natural" paths in curved four-dimensional space-time. And so it goes.

Today, physicists find much that's unnatural about the so-called elementary particles that make up the universe. Everything on Earth is made of familiar electrons and protons and neutrons. But all these particles have heavier "cousins."

No one knows why. It seems completely unnatural. When the first of these "extra" particles—the muon—was discovered, physicist I. I. Rabi asked famously: "Who ordered that?"

Physicists are still trying to answer that question.

Indeed, Einstein said that the question he would most like to answer was: Did God have a choice when he made the universe? Did we have to have a muon? Could gravity have been stronger?

In other words, are the laws of nature themselves "natural" features of the world? Or could they be different in some alternate universe?

What's "natural," of course, depends on context. It's natural for an ice cube to melt on your kitchen counter, but not at the North Pole. Diseases that were "natural" to the Europeans who first took over the Americas spread death among native people who had no "natural" resistance.

Even math has both "natural" and "unnatural" numbers. If we stuck to the natural ones (defined as the positive integers), we wouldn't even have subtraction—not to mention algebra and all the rest.

What's all this got to do with genetically modified food?

It's probably sensible to be suspicious of the motives and methods of global agribusiness, to worry about tinkering with evolution and the dangers of loosing unknown genetic hybrids into the wild.

But natural or unnatural isn't really the point. Genetic modification happens all the time. You could even say that falling in love is nature's way of genetically modifying the species. Does it matter whether it's controlled by a mad rush of hormones or by a batch of tailor-made DNA brewed up in a tube?

We evolve, therefore we are. All other plants and animals, too. There's nothing special about this particular point in the history of any species—corn, humans, or dogs. We're all on our way from someplace, going somewhere.

The real question about genetic modification is: Is it safe? Is it good? Are the benefits worth the risk?

Maybe God didn't have a choice in how he directed the evolution of the universe.

But people do.

Apocalypse Soon

Creation and apocalypse. These are the driving themes of both religion and cosmology. How it all began, and how it will all someday end. Or not.

I was first introduced to Dartmouth cosmologist Marcelo Gleiser through his book on beginnings, *The Dancing Universe: From Creation Myths to the Big Bang.* Now the popular Brazilian writer has worked the entwined threads of science and religion into a vision of "the end" that is strangely comforting and inspiring—that gives us cause to celebrate not only the gift of life but also the finality of death.

"Without limits, there is no desire," writes Gleiser in *The Prophet and the Astronomer: A Scientific Journey to the End of Time.* "And without desire there is no creation. Like stars, which generate pressure to survive the crush of gravity, we create to survive the crush of time."

And as good books do, Gleiser's tells you things you thought you already knew in a way that introduces them to you for the first time.

Take the obvious fact, for example, that the sky is falling— even as we speak.

Buzzing about the solar system like electrons around the nucleus of an atom are something like ten thousand asteroids larger than ten kilometers in diameter—roughly the size of my commute between Santa Monica and UCLA; there are roughly a million larger than one kilometer, and 28 million bigger than a football field.

These interplanetary interlopers come to call with distressing regularity, careening out of control like distracted adolescent drivers, nudged out of their usual orbits by some chance gravitational encounter. When they drop in uninvited, we know what to expect.

In 1908, a puny rock a mere thirty meters in diameter crossed our path and exploded over central Siberia with an energy equivalent to one thousand Hiroshima bombs (roughly the size of an average weapon in today's nuclear arsenal). Thousands of square kilometers of forest were flattened; people were knocked off their feet sixty kilometers away. "As far away as California," Gleiser tells us, "the soot from the blast darkened the skies for several weeks."

As for comets, they are too numerous to count—and at an average of a few kilometers in diameter, large enough to trigger a global catastrophe. Remember what happened when comet Shoemaker-Levy crashed into Jupiter in July of 1994, setting off several days of spectacular planetary fireworks.

Of course, space is plenty big and pretty empty. Maybe it won't happen here. Not in our backyards. At least, not any time soon.

And yet, an impact of the Siberia scale can be expected once every hundred years—which makes us nearly due.

As for a Really Big One, we can expect one of those every 10 million years. It's already been 65 million since an asteroid ten kilometers wide crashed into the Gulf of Mexico, sending up a plume of vaporized rock and water halfway to the moon.

"Most parts of the United States and Mexico were completely destroyed in a matter of hours," Gleiser tells us. "The heavens brought hell to Earth."

Dust equivalent to that spewed from a million volcanoes rose into the air, blocking the sun, turning the Earth dark and cold for months. It wasn't long before one of the most successful species ever to walk the Earth—the dinosaurs—were history.

"What killed the dinosaurs can kill us too," writes Gleiser. "Yes, scientists have become the new prophets of doom."

The good news is that with dinos out of the way, mammals flourished, leading, eventually, to us. Extinction is not all doom and gloom. It's always an opportunity for someone.

But if we wish to survive (and we don't self-destruct first), we could do worse than listen to these prophets with telescopes. Gleiser is not unaware of the irony. "What was first confined to ancient religious texts, prophecies of fiery rocks falling from the skies and bringing widespread destruction, has become a legitimate branch of astronomy."

In an earlier age, the fireworks of Shoemaker-Levy would certainly have been taken as a sign: "Repent or else!" Today, the lesson is: "Prepare or else!" "We have the chance to do something," writes Gleiser, "and we should."

Even when the odds are small, things happen. People win lotteries. "It would be quite foolish to rest the future of civilization (at least of countless lives) on the feeble assurance of small odds," writes Gleiser. "It is a matter not of whether a serious collision will happen but of when."

Eventually, of course, even hell will freeze over; that much we know. The stars will all go out. Perhaps they will be reborn again in another space and time. In the meantime, as Gleiser reminds us, "the skies are full of magic."

Redemption lies in looking up.

Transparency

This is the kind of good news/bad news story I love because it not only sheds light on scientific fraud; it touches on everything from corporate book-cooking to the Homeland Security Act, the First Amendment, the practice of astronomy, and the everyday act of looking out the window.

First, the bad news: Last year, the Lawrence Berkeley National Laboratory in Berkeley had to retract its previously announced discovery of new elements 116 and 118 when no research group (including the group that originally made the discovery) was able to duplicate the results. Now, the lab's internal investigations have concluded this was not simply a mistake. It was fraud. Someone faked the data, and was fired.

Now, the good news: It's the same as the bad news. This is precisely how science is *supposed* to work. Because it is transparent, cheaters get caught. (For the record, the accused scientist filed a grievance against the lab for wrongful termination.)

Why wasn't the fraud caught sooner? Because transparency is not a simple matter.

Take the clothes you're wearing. Are they transparent? You'd probably say no, unless you came to the breakfast table wrapped in cellophane. But while your pj's may be opaque to the prying eyes of your neighbors, they are quite transparent to radio signals, cosmic rays, or neutrinos. (Even a light-year's worth of solid lead is transparent to neutrinos.)

At the same time, your kitchen window, although it appears transparent, is, in fact, opaque to ultraviolet light—which is why you can't get a suntan indoors.

Even on a clear day, the air itself is transparent only to a narrow range of light—visible wavelengths, radio, and a bit of infrared and ultraviolet. To see the full range of electromagnetic radiation, telescopes have to go into space, where they can "see" gamma rays, X rays, and the rest.

(It's a good thing that the atmosphere's ozone layer—or what's left of it—is opaque to most UV radiation: If it weren't, we'd all be fried.)

Only gravity manages to see through everything in the universe. Nothing can hide from it, which is why gravity wave telescopes hold such great promise. Just as gravity reaches inside your clothes to pull on your underwear, so its signals travel to us undisturbed from the very beginning of time, and can tell us what happened in some detail even during the Big Bang.

Transparency is tricky. It depends on how you're looking, and what you're looking through. The Berkeley scientists got fooled by the same incorrect assumption that leads some people to think they can get a suntan through the window: Something they thought was transparent actually wasn't. In this case, it was the workings of their own group.

"There was extreme reliance on one individual because he was considered to be the world's greatest expert in this area," said deputy lab director Pier Oddone. "He was the heart and soul of putting the experiment together. [Fraud] was the last thing anyone would have expected."

There was more. The results were too "pretty" to ignore, said Oddone. They fit expectations so beautifully that no one believed this wasn't Mother Nature's own artistry at work. If some computer had made a mistake, Oddone said, "you don't expect the Mona Lisa to jump out at you." The results were too wonderful *not* to be true.

The combination of beautiful results and trust in a colleague, Oddone said, led the scientists to be "absolutely seduced."

It's all been embarrassing for Berkeley. But in the end, the very transparency of science—the fact that no result is accepted unless and until it is checked (and rechecked) by someone else—is what keeps the enterprise strong.

Just so, democracies protect transparency through everything from a free press to citizens with videocams keeping an eye on cops to guaranteed legal counsel for those accused of breaking the law.

Which is why many people are troubled that the administration's Homeland Security Agency is exempt from both the Freedom of Information Act and basic protection for whistleblowers. Which is also why—despite the threat of terrorism—people are uneasy with secret military tribunals and erosion of civil rights.

At the same time, ordinary American investors—like the Berkeley scientists—allowed themselves to be seduced by wishful thinking and wonderful results: a soaring stock market in a booming economy. We wanted to believe that corporate profits were honestly reported, and so we did. It never occurred to us that things could be so crooked right in our own backyards.

Naked truth isn't always pretty, and glass houses are uncomfortable for everyone. Even so, the price of strength is often just such transparency. It's worth a lot more than meets the eye.

Oops

Oops.

It's one of those worrisome sounds you don't want to hear during surgery or as your plane takes off.

Scientists and newspaper reporters tend to be especially oops-averse because their work is valued precisely for its reliability.

But as we all know, oops happens. And it's not necessarily a bad thing. "Science is inconceivable without mistakes," a physicist friend said to me recently. "You have to push things until they break down. It's the same with people: If we don't push ourselves, we don't know what our limits are."

Then she added: "Being right stokes your ego, but being wrong teaches you something."

She was telling me this in part because I was agonizing over two mistakes that slipped into a recent column, one entirely mine, one the inadvertent result of editing. Both were instructive.

In the first case, I'd written that while light can't escape from a black hole, "gravity waves can and do." I knew it was stretching the point unacceptably when I wrote it (gravity waves are emitted only from the edges of black holes), but I liked the way it sounded, and I told myself I'd go back and modify it later.

Lesson No. 1: If you don't correct mistakes immediately, they have a way of settling in until even the most egregious errors begin to seem maybe-just-okay-after-all.

In the second case, the correct application of an editing rule that changed a "like" to a "such as" altered the sense so that the statement, in context, was wrong.

Lesson No. 2: Even good rules can have bad outcomes.

On the same day, I'd received an e-mail from a friend agonizing that he'd made a fool of himself by mistakenly believing he could restart a failed romantic relationship. I told him I was proud that he would take such risks. As a scientist, he knew it was true. There's no progress without risk, and if you don't make mistakes, chances are, you're not risking enough.

What better time than this—my last "Mind Over Matter" column—to revisit some of my favorite personal oopsies, and what I (should have) learned from them.

My all-time favorite was the time I wrote a long article about scientific fraud and used the deliciously ambiguous example of the Millikan oil drop experiment. Robert Millikan was the physicist who founded Caltech. But with "fraud" resonating in mind, I referred to him throughout the piece as Michael Millikan—a twisted hybrid of Caltech's founder and junk bond king Michael Milken, who pleaded guilty to fraud charges in 1990. (The one person who really seemed to enjoy this "oops" was Millikan's grandson—whose name really is Michael.)

Next in line probably is the time I referred to a subatomic particle called "tau" as "tao." (Tao of physics on the brain, no doubt.)

Lesson No. 3: Even when you think you're concentrating on the task at hand, your brain may be pulling ideas in from entirely different contexts.

In other words: You don't always know what you're thinking (witness the Freudian slip).

My most humiliating mistake was the column I wrote about how easy it is to fool ourselves by listening to higher authorities—only to make an error because I listened to a higher authority. I stated that an article on tabletop fusion appeared on the cover of the journal *Science* because someone "high up" told me it had. Actually, it appeared inside.

Lesson No. 4: Simply being aware of pitfalls doesn't make you immune from them.

I also felt pretty silly a few weeks ago when I described the sun as a star "93 million light-years away," only to have an editor save me. ("Er, did you really mean light-years?")

Lesson No. 5: Difficult tasks take concentration. The routine seems so simple that we don't think twice when we most need to.

Sometimes, of course, a mistake is a matter of perspective. After all, it was certainly an "oops" moment for the dinosaurs when that meteor hit Mexico, but it was pure serendipity for mammals like us. Genetic "mistakes" produce genius and beauty as well as disease. Alexander Fleming mistakenly let a foreign mold waft in through his open window and discovered penicillin.

And Einstein's most famous mistake, as I've mentioned so often in these pages before, may have turned out to be a major discovery. Unaware that the universe was expanding, he inserted a term for a "repulsive force" into his equations to prevent the universe from collapsing. These days, cosmologists think this strange force may account for most of the energy in the cosmos.

This inescapable "oops factor" is wisely acknowledge in the "checks and balances" built into our system of government. If the president makes mistakes, the congress is supposed to correct them (and vice versa).

Unfortunately, it's not recognized in policies like "zero tolerance" or the death penalty. By definition, they don't allow for error.

Oops.

Index